STATISTICAL MECHANICS

R.K. SRIVASTAVA
Formerly, Professor of Physics
Ranchi College, Ranchi University

J. ASHOK
Head, Department of Physics
Ranchi College, Ranchi University

PHI Learning Private Limited
Delhi-110092
2022

₹ 450.00

STATISTICAL MECHANICS
R.K. Srivastava and J. Ashok

ISBN-978-81-203-2782-5 (Print Book)
ISBN-978-93-5443-037-4 (e-Book)

Published by Asoke K. Ghosh, PHI Learning Private Limited, Rimjhim House, 111, Patparganj Industrial Estate, Delhi-110092 and Printed by Syndicate Binders, A-20, Hosiery Complex, Noida, Phase-II Extension, Noida-201305 (N.C.R. Delhi).

Contents

Preface

The subject matter of the book has evolved out of courses given to B.Sc. (Hons.) and M.Sc. students over the years. The book introduces the basic concepts and techniques of Statistical Mechanics to the beginners. We felt that a book on the subject catering to the needs of our university students is needed. We have tried to present the development of the subject in a brief but clear manner. A working knowledge of classical mechanics and quantum mechanics is presupposed.

The basic theory and applications of Statistical Mechanics have found use in diverse fields of study. We have developed the subject following the method of Gibbs, which we consider being logical and systematic, though many simplified newer methods are available. The initial chapters deal with the equilibrium statistical mechanics as applied to both the classical and the quantal model of matter. Towards the end, we discuss some special topics of wide interest, like critical phenomena, Ising model, liquid helium and some simple, selected topics of non-equilibrium statistical mechanics. The graphs used in the text are not to scale. They only illustrate their nature.

We have used a large number of books and journals for source material. We acknowledge some of them in the general reference.

Suggestions for improvement of the text are welcome.

R.K. Srivastava

J. Ashok

Acknowledgements

We are greatly indebted to all our teachers, and express our gratitude to them. Dr. R.K. Srivastava expresses his gratitude to Prof. B. Kumar, Retired Professor of Physics, Ranchi University, Ranchi, who shared his exceptional understanding of the subject with him in their close association over the years.

In the teaching of the subject over a long period, we have consulted a large number of books and journals. We are indebted to all of them, and acknowledge some of them in the general references. We have been greatly influenced by the excellent book of Landau and Lifshitz, and express our gratefulness.

We also wish to thank all the people at Prentice-Hall of India for their meticulous care, cooperation and help in the preparation of the book.

R.K. Srivastava

J. Ashok

Notations Used in the Book

Name	Symbol
Absolute activity	z'
Absolute temperature	T
Area	$\vec{s}$
Beta	$\beta = 1/kT$
Boltzmann constant	k
Boltzmann H-function	H
Canonical partition function	Q
Capacitance	C
Characteristic temperature	θ
Chemical potential	μ
Cluster integral	b_j
Coefficient of viscosity	η
Configurational partition function	$\mathcal{Z}$
Coordinate vector	$\vec{r}$
Correlation function	$C(t)$
Critical temperature	T_c
Current density	j or J
Degeneracy factor	g
Dielectric constant	ε
Diffusion coefficient	D
Dirac delta function	$\delta(x)$
Distribution function	f
Electromotive force	$\mathcal{E}$
Electronic charge	e

Name	Symbol
Energy	$E,\ \varepsilon$
Ensemble average	$\langle\ \rangle$
Enthalpy	W
Entropy	S
Fermi energy	ε_F
Frequency	ν
Fugacity	z
Gamma function	$\Gamma(k)$
Generalized coordinate	q
Generalized momentum	p
Gibbs free energy	G
Grand canonical partition function	Ξ
Grand potential	Ω
Gravitational constant	G
Hamiltonian	H
Heat capacities	C_v and C_p
Helmholtz free energy	A
Impedance	$Z(w)$
Irreducible cluster integral	β_k
Isothermal compressibility	K_T
Kroneckor delta	δ_{ij}
Liouville operator	$\hat{L}$
Magnetic field	$\vec{B}$
Magnetisation	M
Mass of a particle	m
Mayer function	f_{ij}
Nabla operator	$\nabla = \vec{i}\frac{\partial}{\partial x} + \vec{j}\frac{\partial}{\partial y} + \vec{k}\frac{\partial}{\partial z}$
Number density of particles	$\rho = N/V$
Number of particles in a system	N
Operator corresponding to a quantity L	$\hat{L}$
Planck constant	h
Potential energy of interaction between two particles	$u(r_{ij}) = u_{ij}$
Pressure	P
Probability	w and P
Quantity and heat	Q

Name	Symbol
Radial correlation function	$g(r)$
Radial distribution function	$g(r)$
Resistance	R
Single particle partition function	q
Spectral density of random variable	$G(\omega)$
Spin variable	s_i
Statistical matrix	w_{mn}
Statistical operator	$\hat{w}$
Statistical weight	$\Delta\Gamma$
Stefan–Boltzmann constant	σ
Step function	$S(r_{ij})$
Susceptibility	χ
Thermal de broglie wavelength	λ
Temperature	T
Universal gas constant	R
Velocity of light	c
Velocity of longitudinal and transverse elastic waves	c_l and c_t
Virial coefficients	B_k
Volume	V
Volume per particle	$v = V/N$
Wave function	Ψ and ψ
Wave length	λ
Wave vector	$\vec{k}$
Work function	ϕ
Zeta function	$\zeta(k)$

1 Fundamentals of Classical Statistical Mechanics

INTRODUCTION

Statistical mechanics consists of the study of the particular laws which govern the behaviour and properties of macroscopic bodies, that is to say, bodies made up of a very large number of separate particles (atoms and molecules). The number of particles is of the order of 10^{23}, the Avogadro number. From microscopic point of view, complete information about the system can be obtained by setting up the equations of motion and then solving them. In general, the equations of motion are second order differential equations in both classical and quantal case (or twice the number of first order differential equations in Hamiltonian formulation). Their solutions will have two constants of integration for each equation of motion, corresponding to each particle, to be determined from the initial conditions of individual particles. This is a tremendous task involving a lot of paperwork and is a near impossibility. But a new type of regularity can be found by relating the macroscopic description to a few variables based on statistical laws, which arise due to the large number of particles involved.

These laws become quite meaningless if applied to a single particle or even a mechanical system with very few degrees of freedom. The development and exposition of these laws constitute the subject matter of statistical mechanics, which arise due to the large number of particles involved.

The importance of statistical mechanics lies in the fact that, in nature, we are dealing all the time with macroscopic bodies (matter in a bulk) whose behaviour cannot be described by pure mechanical methods and which in fact do obey statistical laws.

Statistical mechanics is divided into two parts: the classical statistical mechanics and the quantal statistical mechanics. If the laws of classical physics are adequate to describe the system, the classical model is chosen. If it becomes necessary to consider the quantal laws of motion, the quantal model is chosen. It is further divided into two parts: the equilibrium statistical mechanics and the non-equilibrium statistical mechanics, dealing with the equilibrium states and the non-equilibrium states, the probabilities and the mean values of the attributes being time-independent or time-dependent in the two cases.

The development of the subject is based on the *Gibbs method*—a general and consistent method.

PHASE SPACE (Γ SPACE)

Consider a system of N identical independent particles (atoms or molecules). The state of such a system is completely specified by $3N$ generalized coordinates and $3N$ generalized momenta. If we conceptualize a $6N$ dimensional space (this is a mathematical concept and cannot be realized in practice), a point in such a space will have $6N$ coordinates. Thus, if we construct a space whose $6N$ axes are the $3N$ generalized coordinates and the corresponding $3N$ generalized momenta, the state of the system represented by $3N$ generalized coordinates and $3N$ generalized momenta will correspond to a point in such a space. This space is called *phase space* or Γ *space*. If all particles are not independent, the system having s degrees of freedom will be described by s generalized coordinates and s corresponding momenta. The phase space in such a case will be $2s$ dimensional space. The state of the system depicted by a point in phase space is called *phase point*. The phase space of a single particle will be 6 dimensional and is referred to as μ *space*.

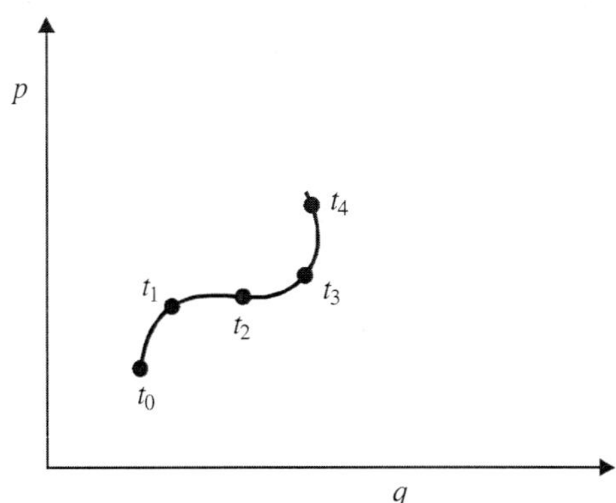

Figure 1.1 Time evolution of phase space—phase trajectory.

With time, the state of the system changes; the generalized $2s$ coordinates take on different sets of values, each state being described by a different phase point. Thus, these phase points will trace a line in phase space, which is called *phase trajectory* (Fig. 1.1).

ENSEMBLE

If we observe a system for a long interval of time, recording its state at a number of equal intervals (the number tending to infinity) of time, we will have a collection of phase points. We consider each of these phase points representing an identical system with the same macroscopic properties but of different microscopic states. We thus have a large number of systems (the number tending to infinity) similar in structure but suitably randomized in microstates. This collection of imagined systems, which are similar, independent, and non-interacting is called *ensemble*. Gibbs replaced the time-dependent picture by a static (time-independent) picture through the concept of ensemble. Instead of contemplating directly the succession of phases through which the system passes during the course of time, we use the ensemble, which is a system of systems, each of which is executing the same motion but in a different phase at any given instant of time, and statistical features can be studied by surveying the ensemble. The average over all the elements in an ensemble is called *ensemble average*. We shall use ensemble average to replace time average over all the states of a single system.

DISTRIBUTION FUNCTION

A macroscopic body that does not interact with other bodies is called a *closed system*. A small part of it, which is still macroscopic, is called a *subsystem*. A subsystem is not closed; it undergoes all kinds of interactions with other parts of the system. Because of the large number of degrees of freedom, the interactions are very complicated and the state of the subsystem will change with time in a very complicated and involved way.

The basis for the statistical approach is the fact that owing to the complex nature of interactions, the subsystem will pass sufficiently, often through all its possible states. If we denote by $\Delta p\ \Delta q$ ($\equiv \Delta p_1 \ldots \Delta p_s, \Delta q_1 \ldots \Delta q_s$, s degrees of freedom) a small region of phase space lying between the interval q_1 and $q_1 + \Delta q_1, \ldots, q_s$ and $q_s + \Delta q_s$, p_1 and $p_1 + \Delta p_1, \ldots,$ p_s and $p_s + \Delta p_s$, and let Δt be the part of a sufficiently large interval of time T, during which the phase point was in the cell of volume $\Delta p\ \Delta q$. The fraction $\Delta t/T$ tends to a limit, and is called probability w.

$$w = \underset{T\to\infty}{\text{Lt}} \frac{\Delta t}{T} \tag{1.1}$$

This represents the probability that the subsystem will be found in the given cell, if observed at any arbitrary time.

We define a function $f(p, q) = f(p_1 \ldots p_s, q_1 \ldots q_s)$, so that $f(p, q)\ dp\ dq$ becomes the probability dw of states described by points in this phase element, such that

$$dw = f(p, q)\ dp\ dq \tag{1.2}$$

where

$$dp\ dq = dp_1\ dp_2 \ldots dp_s\ dq_1\ dq_2 \ldots dq_s \tag{1.3}$$

$f(p, q)$ is a function of all the coordinates and all the momenta, and is called *statistical distribution function* or simply *distribution function* or *probability density*. The distribution function obviously satisfies the normalisation condition (i.e., must integrate to unity).

$$\int f(p, q)\ dp\ dq = 1 \tag{1.4}$$

This stresses the fact that the sum of probabilities of all possible states is unity.

The time dependence of distribution function will be discussed later in connection with Liouville theorem.

The determination of statistical distribution function for any subsystem is the basic problem of statistical mechanics.

The mean value of any quantity $L(p, q)$ that depends on the state of this subsystem, is given by

$$\langle L \rangle = \int_{\text{all states}} L(p,q)\, f(p,q)\, dp\, dq \tag{1.5}$$

The suffix 'all states' means that we multiply all possible values of $L(p, q)$ with their corresponding probabilities and integrate over all states.

Averaging by means of distribution function is called *statistical averaging* or *ensemble averaging*.

The time averaging will give

$$\langle L \rangle = \underset{T \to \infty}{\text{Lt}} \frac{1}{T} \int L(p,q)\, dt \tag{1.6}$$

For this, we must solve the equations of motion using the initial conditions of p and q, substitute and calculate $L(p, q)$ and then use Eq. (1.6). The solution of equations of motion is practically impossible because of large number of degrees of freedom, as we have already pointed out. For this reason, Gibbs devised the less direct method of ensemble averaging.

We assume that statistical averaging is completely equivalent to time averaging. This is known as *ergodic hypothesis*. This may not be rigorously true but odds are overwhelmingly in its favour. We assume the validity of the equivalence without attempting a justification.

It is apparent that the deductions made about macroscopic bodies from statistical mechanics are probabilistic in contrast to the deductions made from mechanics, that are deterministic. This is so because the results depend on a much smaller data than a complete mechanical description would require. However, the probabilistic nature of statistical mechanics does not appear when we apply it in practice to macroscopic bodies. The reason for it is that, if we observe the macroscopic body for a large interval of time, the physical quantities (macroscopic properties) remain practically constant and equal to their mean values. If a closed system is in such a state that for any macroscopic subsystem all macroscopic physical quantities approximate very closely to their mean values, the system is said to be in a state of *statistical equilibrium*. It is evident that if a closed system is observed for a sufficiently long time, for a greater part of it, it will be in statistical equilibrium. If it is temporarily disturbed by some external influence and afterwards again becomes a closed system, then eventually it must return to a state of statistical equilibrium. The period of time in which transition to statistical equilibrium occurs is called *relaxation time*.

In practice, three types of ensembles are useful. *Microcanonical ensemble* represents closed (or isolated) systems so that no exchange of energy E or mass (i.e., number of particles) occurs between them. In other words E, N and T are taken as constants. *Canonical ensemble* represents isothermal systems, which can exchange energy with one another but not mass. In this case, N and temperature T remain constant. *Grand canonical ensemble* represents open isothermal systems, which can exchange both E and N. In this case, T and chemical potential μ (chemical potential will be discussed later) remain constant. Other types of ensemble are possible, but are not in general use.

LIOUVILLE THEOREM

The Liouville theorem can be stated as: *The distribution function is constant along a phase trajectory of a subsystem.* Suppose that the distribution function with its time dependence is known and is represented by $f(p, q, t)$. We can then write

$$\frac{df}{dt} = \frac{\partial f}{\partial t} + \sum_i \frac{\partial f}{\partial p_i}\frac{dp_i}{dt} + \sum_i \frac{\partial f}{\partial q_i}\frac{dq_i}{dt} \tag{1.7}$$

Consider an arbitrary but fixed volume V in phase space of surface $\vec{s}$. Then the number of phase points in the volume V at time t is

$$N_V = N\int_V f(p,q,t)\, dp\, dq \tag{1.8}$$

Consequently,

$$\frac{dN_V}{dt} = N\int_V \frac{\partial f(p,q,t)}{\partial t}\, dp\, dq \tag{1.9}$$

The number of phase points moving across the surface $\vec{s}$ per unit time is also given by

$$\frac{dN_V}{dt} = -N\int_{\vec{s}} (f\vec{v})\cdot d\vec{s} = -N\int_V \text{div}\,(f\vec{v})\,dV \tag{1.10}$$

where $\vec{v}$ is the velocity.

Comparison of Eq. (1.10) with Eq. (1.9) gives

$$\frac{\partial f}{\partial t} = -\text{div}(f\vec{v}) = -\sum_i \left(\frac{\partial f}{\partial p_i}\dot{p}_i + \frac{\partial f}{\partial q_i}\dot{q}_i\right) - \sum_i f\left(\frac{\partial \dot{p}_i}{\partial p_i} + \frac{\partial \dot{q}_i}{\partial q_i}\right)$$

since $f\vec{v}$ has components $f\dot{p}_1, \ldots, f\dot{p}_s, f\dot{q}_1, \ldots, f\dot{q}_s$.

The last term in the above expression vanishes on account of Hamilton equations of motion.

$$\dot{q}_i = \frac{\partial H}{\partial p_i},\quad \dot{p}_i = -\frac{\partial H}{\partial q_i}$$

where H is the Hamiltonian function, giving us

$$\frac{\partial f}{\partial t} = -\sum_i \left(\frac{\partial f}{\partial p_i}\dot{p}_i + \frac{\partial f}{\partial q_i}\dot{q}_i\right) \tag{1.11}$$

or

$$\frac{\partial f}{\partial t} = -\sum_i \left(-\frac{\partial f}{\partial p_i}\frac{\partial H}{\partial q_i} + \frac{\partial f}{\partial q_i}\frac{\partial f}{\partial p_i} \right) = -\{f, H\} \quad (1.12)$$

where $\{f, H\}$ is Poisson bracket of f and H.

Substitution of Eq. (1.11) in Eq. (1.7) gives

$$\frac{df}{dt} = 0 \quad (1.13)$$

That is, the probability density is constant, or in other words, the 'probability fluid' is incompressible. Thus, the distribution function is independent of time and hence no crowding of the phase points in some favoured region of phase space occurs.

Liouville theorem is valid if we are dealing with closed systems or at best with quasi-closed (almost closed) systems for not too long intervals of time, during which the system behaves like a closed system with sufficient accuracy.

POSTULATE OF CLASSICAL STATISTICAL MECHANICS

At this stage we introduce the fundamental postulate of classical statistical mechanics known as *the postulate of equal a priory probability. It states that for a macroscopic system in equilibrium, all different microstates conforming to the macroscopic conditions of the system are equally probable.*

The macroscopic state of the system is characterized by the variables E, N and V. A very large number of different microscopic states will satisfy the said macroscopic state. The postulate states that all of them have equal probability, i.e. no microscopic state has preference of one over the other. This law is the basis of statistical mechanics. There is no direct proof for the law, but it appears reasonable, does not contradict any known law of mechanics, and leads to results which agree with the observations.

MICROCANONICAL DISTRIBUTION

Let us consider a closed system in statistical equilibrium. We start with the basics of the theory of probability: The probability $f_{12}\, dp_1\, dp_2\, dq_1\, dq_2$ of combined system consisting of two statistically independent subsystems 1 and 2, is equal to the product of separate probabilities $f_1\, dp_1\, dq_1$ and $f_2\, dp_2\, dq_2$ of the two subsystems. That is,

$$f_{12}\, dp_1\, dp_2\, dq_1\, dq_2 = f_1\, dp_1\, dq_1\, f_2\, dp_2\, dq_2$$

or

$$\ln f_{12} = \ln f_1 + \ln f_2 \quad (1.14)$$

This suggests that logarithm of the distribution function is additive. We also infer from Liouville theorem that the distribution function remains constant while the subsystem moves like a closed system. This suggests that f can be expressed only as a function of integrals of motion. Combining these two inferences, we conclude that logarithm of the distribution function must be the function of additive integrals of motion.

From classical mechanics we know that there are only seven independent additive integrals of motion: the energy, the three components of the momentum, and the three components of the angular momentum. We thus reach a very important conclusion that only the seven independent additive integrals of motion completely define the statistical distribution function of a closed subsystem, and therefore, the mean values of any physical quantity related to it. We can further exclude the momentum and the angular momentum from the consideration, if we imagine the system to be enclosed in a rigid box and consider a system of coordinates in which the box is at rest. The momentum and the angular momentum then cease to be integrals of motion and only the energy remains the additive integral of motion. The presence of box does not affect the statistical properties of the subsystem. We thus have

$$\ln f_a = \alpha_a + \beta E_a(p, q) \tag{1.15}$$

where α and β are constants and the suffix a refers to the subsystem 'a'.

We can now set a simple distribution function for a closed system. The simplest such function is

$$\begin{aligned} f &= \text{constant} \qquad \text{for } E(p, q) = E_0 \\ &= 0 \qquad\qquad \text{otherwise} \end{aligned} \tag{1.16}$$

For the integral $\int f(p, q)\, dp\, dq$ to be non-zero, the function $f(p, q)$ must be infinite at the point $E(p, q) = E_0$. This gives

$$f = \text{constant } \delta(E - E_0) \tag{1.17}$$

where the δ is the Dirac-delta function defined by

$$\begin{aligned} \delta(x - x_0) &= \infty \qquad \text{if } x = x_0 \\ \delta(x - x_0) &= 0 \qquad \text{if } x \neq x_0 \end{aligned} \tag{1.18}$$

and

$$\int_{-\infty}^{\infty} \delta(x - x_0)\, dx = 1 \tag{1.19}$$

The distribution given by Eq. (1.17) is called the *microcanonical distribution.*

STATISTICAL WEIGHT AND ENTROPY

Let $\Delta p \Delta q$ characterize the volume of the region of phase space in which the given subsystem remains nearly all the time. Then,

$$f(\langle E \rangle)\, \Delta p\, \Delta q = 1 \tag{1.20}$$

where the distribution function has been expressed as a function of mean energy.

We define a quantity $\Delta\Gamma$, called statistical weight, by

$$\Delta\Gamma = \frac{\Delta p \Delta q}{(2\pi\hbar)^s} \tag{1.21}$$

where s is the number of degrees of freedom and $\hbar = h/2\pi$, h being Planck constant.

Replacing $d\Gamma_a/dE_a$ by $\Delta\Gamma_a/\Delta E_a$, where ΔE_a is the energy interval corresponding to $\Delta\Gamma_a$, and using Eqs. (1.22) and (1.26), we get

$$f(E)dp\,dq = \text{constant } \delta(E - E_0)e^{\sigma}\prod_a dE_a \tag{1.27}$$

where we have absorbed the factor ΔE_a in the constant because the dependence of ΔE_a on energy is negligible.

Now let the closed system be not in a state of equilibrium. Then, its macroscopic state will vary with time until the system eventually reaches the state of equilibrium. The successive states through which the system passes will then correspond to the energy distributions of successive greater probabilities. The probabilities, as given by Eq. (1.27), then suggest that successive states have higher values of σ. Thus the processes, taking place in a closed system not in equilibrium, are such that the system passes from the states of lower value of σ to the states of higher value of σ until σ attains its maximum value, corresponding to the maximum value of probability at statistical equilibrium. This establishes the law of increase of σ.

To relate σ to the pressure P and temperature T, let us consider two subsystems forming a closed system. The energy E of the system will be the sum of energies E_1 and E_2 of the two subsystems and the same applies to σ (it is additive), i.e.,

$$E = E_1 + E_2 \tag{1.28}$$

$$\sigma = \sigma_1 + \sigma_2$$

Since E is constant, we have $dE_1 = -dE_2$ and as the volume is constant, we have

$$\left(\frac{\partial \sigma}{\partial E_1}\right)_V = \left(\frac{\partial \sigma_1}{\partial E_1}\right)_V - \left(\frac{\partial \sigma_2}{\partial E_2}\right)_V$$

If we write

$$\left(\frac{\partial \sigma_1}{\partial E_1}\right)_V = \frac{1}{\Theta_1}, \quad \left(\frac{\partial \sigma_2}{\partial E_2}\right)_V = \frac{1}{\Theta_2} \tag{1.29}$$

then

$$\left(\frac{\partial \sigma}{\partial E_1}\right)_V = \frac{1}{\Theta_1} - \frac{1}{\Theta_2}$$

We see from the above, that at equilibrium, σ is maximum and $d\sigma/dE_1 = 0$. This is satisfied if $\Theta_1 = \Theta_2$. Also from the law of increase of σ, $d\sigma$ is always positive. Hence dE_1 is positive if $\Theta_1 < \Theta_2$, i.e. energy will flow from the second part to the first part. Thus, Θ has the attributes of temperature but the dimension of energy.

We can then write

$$\Theta = kT \tag{1.30}$$

where k is Boltzmann constant, and

$$\left(\frac{\partial \sigma}{\partial E}\right)_V = \frac{1}{\Theta} \tag{1.31}$$

Alternatively,

$$k\left(\frac{\partial \sigma}{\partial E}\right)_V = \frac{1}{T} \tag{1.32}$$

In thermal equilibrium, the entropy of a system for a given value of energy depends upon its volume but not on the shape (true only for gases and liquids; in solids, the change of shape needs work done and hence expenditure of energy). The force acting on an area $d\vec{s}$ of the surface bounding its volume with $\vec{r}$ as the radius vector of the element of area, taking energy E as a function of S and $\vec{r}$, we have from mechanics

$$\vec{F} = -\left(\frac{\partial E}{\partial \vec{r}}\right)_S \tag{1.33}$$

Using the change in volume $\partial V = \partial \vec{s} \cdot \partial \vec{r}$, we get

$$\vec{F} = -\left(\frac{\partial E}{\partial V}\right)_S d\vec{s} \tag{1.34}$$

and the force per unit area or pressure P is given by

$$P = -\left(\frac{\partial E}{\partial V}\right)_S \tag{1.35}$$

Equations (1.32) and (1.35) then yield

$$k\left(\frac{\partial \sigma}{\partial V}\right)_E = -\frac{P}{T} \tag{1.36}$$

We have thus shown that σ is a dimensionless quantity and has all the attributes of entropy. Hence we write

$$\sigma = \frac{S}{k} \tag{1.37}$$

THERMODYNAMIC POTENTIALS

Thermodynamics is a phenomenological theory of heat. The physical system is described by means of a limited number of macroscopically measurable quantities. We now review a number of thermodynamic relations between such quantities. We have already done so for two of them. Rewriting, we have

$$T = \left(\frac{\partial E}{\partial S}\right)_V \tag{1.38}$$

and

$$P = -\left(\frac{\partial E}{\partial V}\right)_S \tag{1.39}$$

In a body not thermally isolated, energy can be acquired by means of transmission from other bodies as well as through work. If we take as positive, dQ the heat received by the body and dR the work done on it, then from the principle of conservation of energy

$$dE = dQ + dR \tag{1.40}$$

The above is the first law of thermodynamics, which modifies to

$$dE = dQ - P\,dV \tag{1.41}$$

where $dR = -P\,dV$, the work done by the body in expansion at constant pressure.

From the second law of thermodynamics, we have

$$dQ = T\,dS \tag{1.42}$$

Equations (1.41) and (1.42) are combined to yield

$$dE = T\,dS - P\,dV \tag{1.43}$$

Thus, E is a function of S and V. Equations (1.38) and (1.39) can easily be seen to follow from it.

The specific heats are

$$C_V = \left(\frac{\partial Q}{\partial T}\right)_V = T\left(\frac{\partial S}{\partial T}\right)_V \tag{1.44}$$

and

$$C_P = \left(\frac{\partial Q}{\partial T}\right)_P = T\left(\frac{\partial S}{\partial T}\right)_P \tag{1.45}$$

The four Maxwell relations are

$$\left(\frac{\partial S}{\partial V}\right)_T = \left(\frac{\partial P}{\partial T}\right)_V, \quad \left(\frac{\partial S}{\partial P}\right)_T = -\left(\frac{\partial V}{\partial T}\right)_P$$

$$\left(\frac{\partial T}{\partial P}\right)_S = \left(\frac{\partial V}{\partial S}\right)_P, \quad \left(\frac{\partial T}{\partial V}\right)_S = -\left(\frac{\partial P}{\partial S}\right)_V \tag{1.46}$$

In addition to the above, to describe the behaviour of a system at equilibrium, we use certain energy terms defined under various constraints, which are called *thermodynamic potentials*.

The first thermodynamic potential of a system is its energy, which we have already defined by Eq. (1.43). The next thermodynamic potential we define is enthalpy or heat content or heat function. We know that in throttling process, a quantity $W = E + PV$ remains constant. This is called *enthalpy*, i.e.,

$$W = E + PV \tag{1.47}$$

and for infinitesimal changes in it

$$dW = d(E + PV) = T\,dS + V\,dP \tag{1.48}$$

in which Eq. (1.43) has been used. Thus, W is a function of S and P. And hence

$$T = \left(\frac{\partial W}{\partial S}\right)_P, \quad V = \left(\frac{\partial W}{\partial P}\right)_S \tag{1.49}$$

An isothermal process is always irreversible and hence entropy always increases. In adiabatic process, entropy remains constant. These can be combined to yield

$$\frac{dQ}{T} \le dS \tag{1.50}$$

If T remains constant, then we have

$$dQ \le d(TS) \tag{1.51}$$

Use of Eq. (1.41) yields

$$P\,dV \le -d(E - TS) \tag{1.52}$$

If the volume remains constant, left hand vanishes and we reach a theorem that in isothermal isovolumetric process, the quantity $E - TS = A$ (say), called *Helmholtz free energy*, is minimum in the equilibrium state. We have thus defined the Helmholtz free energy by

$$A = E - TS \tag{1.53}$$

and for infinitesimal changes in it,

$$dA = dE - T\,dS - S\,dT$$

which reduces, in view of Eq. (1.43), to

$$dA = -P\,dV - S\,dT \tag{1.54}$$

from which follows

$$P = -\left(\frac{\partial A}{\partial V}\right)_T \tag{1.55}$$

and

$$S = -\left(\frac{\partial A}{\partial T}\right)_V \tag{1.56}$$

Thus, A is a function of V and T.

If the process is isothermal isobaric, we then have from Eqs. (1.51) and (1.41)

$$0 \leq -d(E - TS + PV) \tag{1.57}$$

We reach a theorem that in isothermal isobaric process, the quantity $(E - TS + PV) = G$ (say), called *Gibbs free energy*, is minimum in the state of thermal equilibrium. We have thus defined the Gibbs free energy by

$$G = E - TS + PV = A + PV \tag{1.58}$$

and for infinitesimal changes in it, with the aid of Eq. (1.43), we have

$$dG = V\mathrm{d}P - S\mathrm{d}T \tag{1.59}$$

and whence

$$V = \left(\frac{\partial G}{\partial P}\right)_T \tag{1.60}$$

and

$$S = -\left(\frac{\partial G}{\partial T}\right)_P \tag{1.61}$$

Thus G has been defined as a function of P and T.

Of the thermodynamic variables P, T, V and S, any two of them can be taken as independent variables, the other two are then found from Eq. (1.46). The thermodynamic potentials E, W, A and G have been defined above as functions of two of the variables from P, T, V and S. The variables V and S are themselves additive. The additivity of a quantity means that when the amount of the substance is changed, i.e., the number N of the particles is changed, the quantity will change by the same factor. This means that the dependence on N is linear and in the infinitesimal changes of thermodynamic potentials, we must add a term proportional to dN, the infinitesimal change in the number of particles:

$$dE = T\,dS - P\,dV + \mu dN \tag{1.62}$$

$$dW = T\,dS - P\,dP + \mu dN \tag{1.63}$$

$$dA = -S\,dT - P\,dV + \mu dN \tag{1.64}$$

$$dG = -S\,dT - P\,dP + \mu dN \tag{1.65}$$

where the quantity μ is called *chemical potential.*

For an isothermal isobaric process, we have from Eq. (1.65)

$$\mu = \left(\frac{\partial G}{\partial N}\right)_{P,T} \tag{1.66}$$

As G is an additive function of P, T and N, we immediately have

$$G = \mu N$$

or

$$\mu = \frac{G}{N} \tag{1.67}$$

Thus, the chemical potential for a body of identical particles, is Gibbs free energy per particle. We can also write

$$\mu = \left(\frac{\partial E}{\partial N}\right)_{S,V} = \left(\frac{\partial W}{\partial N}\right)_{S,P} = \left(\frac{\partial A}{\partial N}\right)_{V,T} \tag{1.68}$$

From Eqs. (1.65) and (1.67), we can write

$$d\mu = \frac{V}{N}dP - \frac{S}{N}dT \tag{1.69}$$

This is known as *Gibbs-Duhem equation.*

Under the constraint of constant volume but variable number of particles, we define a new thermodynamic potential Ω called grand potential, (on account of its relation with grand canonical partition function), as:

$$\Omega = A - G = A - \mu N = -PV \tag{1.70}$$

and

$$d\Omega = -S\,dT - N\,d\mu \tag{1.71}$$

whence

$$S = -\left(\frac{\partial \Omega}{\partial T}\right)_{\mu,V} \tag{1.72}$$

and

$$N = -\left(\frac{\partial \Omega}{\partial \mu}\right)_{V,T} \tag{1.73}$$

We can show that in a state of thermal equilibrium at constant T, V and μ, the grand potential is minimum.

ENTROPY OF AN IDEAL GAS USING CLASSICAL MICROCANONICAL DISTRIBUTION

Consider an ideal gas of N identical particles (degrees of freedom $s = 3N$) of mass m in a volume V. The statistical weight Γ for energy less than or equal to E is

$$\Gamma = \frac{1}{h^{3N}} \int_{H<E} dp_1 \ldots dp_{3N}\, dq_1 \ldots dq_{3N} \tag{1.74}$$

The integral over q can be immediately carried out to yield V^N. The integration over p is the volume of a $3N$-sphere of radius

$$p = \sqrt{2mE}$$

The volume* of an s-sphere of radius r is

$$\pi^{s/2} \frac{r^s}{(s/2)!}$$

This gives

$$\Gamma = \frac{V^N}{h^{3N}} \frac{\pi^{3N/2}}{(3N/2)!} (2mE)^{3N/2} \tag{1.75}$$

Using Eqs. (1.22) and (1.37), we get for entropy

$$S = Nk \ln\left[V\left(\frac{4\pi mE}{3h^2 N}\right)^{3/2}\right] + \frac{3}{2} Nk \tag{1.76}$$

---

*The volume V_n of an n-sphere is

$$V_n = \int dV_n = \int_{x_1^2+\cdots+x_n^2=r^2} dx_1\, dx_2 \ldots dx_n = C_n r^n$$

where C_n is a constant, evaluated as below. Let

$$I = \int e^{-(x_1^2+\ldots+x_n^2)} dx_1 \ldots dx_n = \int (e^{-x^2} dx)^n = \pi^{n/2}$$

(For integrals of this kind, see first footnote in Chapter 3).
The surface area S_n of the n-sphere is related to its volume V_n by

$$\frac{dV_n}{dr} = S_n$$

Hence

$$I = \int e^{-r^2} S_n dr = nC_n \int r^{n-1} e^{-r^2} dr = \frac{n}{2} C_n \int t^{(n-2)/2} e^{-t}\, dt \quad (\text{where } t = r^2)$$

$$= \frac{n}{2} C_n \Gamma\left(\frac{n}{2} - 1\right) = C_n \left(\frac{n}{2}\right)!$$

This gives

$$C_n = \frac{\pi^{n/2}}{(n/2)!}$$

and hence

$$V_n = \frac{\pi^{n/2}}{(n/2)!} r^n$$

We have used Stirling approximation* $\ln n! = n \ln n - n$ (for large n). Solving for E, this yields

$$E = \frac{3}{4\pi}\frac{h^2}{m}\frac{N}{V^{2/3}}\exp\left(\frac{2}{3}\frac{S}{Nk} - 1\right)$$

Use of Eqs. (1.38), (1.39) and (1.44) gives

$$T = \frac{2}{3}\frac{E}{Nk} \tag{1.77}$$

$$P = \frac{2}{3}\frac{E}{V} = \frac{NkT}{V} \tag{1.78}$$

and

$$C_V = \frac{3}{2}Nk \tag{1.79}$$

These are the well-known results.

Equation (1.76) can be rewritten as

$$S = Nk \ln\left[V\left(\frac{E}{N}\right)^{3/2}\right] + NS_0 \tag{1.80}$$

where

$$S_0 = \tfrac{3}{2}k\left(1 + \ln\frac{4\pi m}{3h^2}\right) \tag{1.81}$$

---

*We have

$$n! = \Gamma(n+1) = \int_0^\infty dt\, t^n e^{-t} = \int_0^\infty dt\, e^{\ln t^n} e^{-t} = \int_0^\infty dt\, e^{(n\ln t - t)}$$

Put $t = n + \sqrt{n}x$ so that $dt = \sqrt{n}\,dx$. This gives

$$n! = \sqrt{n}\int_{-\sqrt{n}}^{\infty} dx\, e^{\left[n\ln(n+\sqrt{n}x) - n - \sqrt{n}x\right]}$$

For large n, the lower limit can be replaced by $-\infty$ and we have

$$\ln(n + \sqrt{n}x) = \ln n + \ln\left(1 + \frac{x}{\sqrt{n}}\right) \simeq \ln n + \frac{x}{\sqrt{n}} - \frac{x^2}{n}$$

This gives

$$n! \simeq \sqrt{n}\, n^n e^{-n}\int_{-\infty}^{\infty} dx\, e^{-x^2/2} = \sqrt{2\pi n}\, n^n e^{-n}$$

Hence

$$\ln n! \simeq \frac{1}{2}\ln 2\pi + \left(n + \frac{1}{2}\right)\ln n - n$$

For large n, we can further approximate to

$$\ln n! \simeq n\ln n - n$$

Now consider two ideal gases, both at the same temperature, having number of particles N_1 and N_2, entropy S_1 and S_2, and volume V_1 and V_2. When the gases are mixed, the volume of the mixture will be $V = V_1 + V_2$, entropy of the mixture is S_{12} (say). The process is irreversible and the increase in entropy is given by

$$\Delta S = S_{12} - (S_1 + S_2) > 0 \tag{1.82}$$

Substituting from Eq. (1.80) and solving, we get

$$\frac{\Delta S}{k} = N_1 \ln \frac{V}{V_1} + N_2 \ln \frac{V}{V_2} > 0 \tag{1.83}$$

Since V is greater than both V_1 and V_2, the left-hand side is always positive, validating the inequality given by Eq. (1.83). This conforms to the experimental observation if the two gases are different. However, if the two gases are same, the result is disastrous. Any existing gas can be arrived at by putting any imaginary number of partitions in it and then taking them off. Since the number of partitions can be any digit, the very definition of entropy as a function of the thermodynamic state of the gas loses its meaning. This is known as *Gibbs paradox*.

Gibbs resolved the problem empirically. He divided the right-hand expression of Eq. (1.75) by a factor $N!$ to yield in place of Eq. (1.80):

$$S = Nk \ln \left[\frac{V}{N} \left(\frac{E}{N} \right)^{3/2} \right] + NS_0 \tag{1.84}$$

with S_0 now given by

$$S_0 = \frac{3}{2} k \left(\frac{5}{3} + \ln \frac{4\pi m}{3h^2} \right) \tag{1.85}$$

The division by $N!$ does not affect the equation of state and other thermodynamic functions. The mixture of two gases will give

$$\frac{\Delta S}{k} = N_1 \ln \frac{V_1 + V_2}{V_1} + N_2 \ln \frac{V_1 + V_2}{V_2} \quad \text{for two different gases} \tag{1.86}$$

and

$$\frac{\Delta S}{k} = (N_1 + N_2) \ln \frac{V_1 + V_2}{N_1 + N_2} - N_1 \ln \frac{V_1}{N_1} - N_2 \ln \frac{V_2}{N_2} \quad \text{for the same gas} \tag{1.87}$$

For a mixture of two different gases, $\Delta S > 0$, on account of V (= $V_1 + V_2$) being greater than both V_1 and V_2.

However, if the two gases are same, the specific volume, i.e., volume per particle, is same, the result becomes

$$\Delta S = 0$$

thereby removing the paradox.

The reason for the factor $N!$ is inherently quantal and has no classical explanation. In the quantal model, the particles are indistinguishable and an interchange of particles does not produce a new state. The N particles will have $N!$ permutations amongst themselves, and hence the counting of states must be divided by the factor $N!$. This is known as 'correct Boltzmann counting' and must be appended to classical calculations to arrive at the proper result. In the classical picture, the particles are distinguishable and we shall continue to regard them as such, except for the counting of number of states.

EXERCISES

1. Explain the following:
   (a) Microstate,
   (b) Macrostate,
   (c) Phase space,
   (d) Statistical distribution function,
   (e) Ensemble,
   (f) Postulate of equal a priori probability.

2. State and prove Liouville theorem. How is it analogous to the equation of continuity of an incompressible fluid?

3. What is statistical weight? How is it related to entropy?

4. What is a microcanonical ensemble? Obtain Gibbs microcanonical distribution.

5. What is Gibbs paradox? How is it resolved?

6. Verify the validity of Liouville theorem for an elastic collision of two particles moving in a straight line.
Hint: Let p_1, q_1; p_2, q_2 and p_1', q_1'; p_2' q_2' represent the momenta and positions of the two particles before and after collision. Then from the conservation of momentum and energy, we have

$$p_1' + p_2' = p_1 + p_2;\ \frac{p_1^2}{2m_1} + \frac{p_2^2}{2m_2} = \frac{p_1'^2}{2m_1} + \frac{p_2'^2}{2m_2}$$

Hence

$$p_1' = \frac{m_1 - m_2}{m_1 + m_2}\, p_1 + \frac{2m_1}{m_1 + m_2}\, p_2$$

$$p_2' = \frac{m_1 - m_2}{m_1 + m_2}\, p_2 + \frac{2m_1}{m_1 + m_2}\, p_1$$

With obvious relation $\partial p_i'/\partial q_j = 0$, $\partial q_i'/\partial q_j = \delta_{ij}$, we find that the Jacobian of transformation $|J| = 1$, i.e. the volume in phase space is constant.

7. Find the phase trajectory of a harmonic oscillator and thus find the phase volume.
Hint: The Hamiltonian is $H = (p^2/2m) + 1/2\ kq^2$; q and p are given by

$$q = a\cos(\omega t + \delta),\ p = -m\omega\, a \sin(\omega t + \delta),\ \omega = \sqrt{k/m}$$

The energy is

$$E = \frac{1}{2} m\omega^2 a^2$$

Eliminating t, we have

$$\frac{p^2}{2mE} + \frac{q^2}{2E/m\omega^2} = 1$$

This is an equation of an ellipse, which represents the phase trajectory. The phase volume is the area of the ellipse = $2\pi E/\omega$.

8. For an ideal gas of N identical particles in microcanonical ensemble, find the entropy, the equation of state, and the specific heat.

9. Explain the thermodynamic potentials and their dependence on the number of particles in the system.

10. Obtain the relation

$$\left(\frac{\partial E}{\partial V}\right)_T = T\left(\frac{\partial P}{\partial T}\right)_V - P$$

Hint: We have $dE = \left(\frac{\partial E}{\partial T}\right)_V dT + \left(\frac{\partial E}{\partial V}\right)_T dV = T\,dS - P\,dV$

$$= T\left(\frac{\partial S}{\partial V}\right)_T dV + T\left(\frac{\partial S}{\partial T}\right)_V dT - P\,dV$$

$$= T\left(\frac{\partial P}{\partial T}\right)_V dV + T\left(\frac{\partial S}{\partial T}\right)_V dT - P\,dV$$

Hence

$$\left(\frac{\partial E}{\partial V}\right)_T = T\left(\frac{\partial P}{\partial T}\right)_V - P$$

11. Prove

$$C_p - C_v = T\left(\frac{\partial V}{\partial T}\right)_P \left(\frac{\partial P}{\partial T}\right)_V = -T\left[\left(\frac{\partial V}{\partial T}\right)_P\right]^2 \left(\frac{\partial P}{\partial V}\right)_T = -T\left[\left(\frac{\partial P}{\partial T}\right)_V\right]^2 \left(\frac{\partial V}{\partial P}\right)_T$$

Hint: From the two $T\,dS$ equations, we have

$$T\,dS = C_V\,dT + T\left(\frac{\partial P}{\partial T}\right)_V dV \text{ and } C_P\,dT - T\left(\frac{\partial V}{\partial T}\right)_P dP = T\,dS$$

This gives

$$(C_P - C_V)\,dT = T\left(\frac{\partial V}{\partial T}\right)_P dP + T\left(\frac{\partial P}{\partial T}\right)_V dV$$

Hence

$$\left(C_P - C_V\right)\left[\left(\frac{\partial T}{\partial V}\right)_P dV + \left(\frac{\partial T}{\partial P}\right)_V dP\right] = T\left(\frac{\partial V}{\partial T}\right)_P dP + T\left(\frac{\partial P}{\partial T}\right)_V dV$$

Composing the coefficient of dP or dV, we get

$$(C_P - C_V) = T\left(\frac{\partial V}{\partial T}\right)_P \cdot \left(\frac{\partial P}{\partial T}\right)_V$$

Use of Maxwell relations gives

$$\left(\frac{\partial P}{\partial T}\right)_V = \left(\frac{\partial S}{\partial V}\right)_T = \left(\frac{\partial S}{\partial P}\right)_T \left(\frac{\partial P}{\partial V}\right)_T = -\left(\frac{\partial V}{\partial T}\right)_P \left(\frac{\partial P}{\partial V}\right)_T$$

$$\therefore \quad C_P - C_V = -T\left[\left(\frac{\partial V}{\partial T}\right)_P\right]^2 \left(\frac{\partial P}{\partial V}\right)_T$$

Similarly,

$$\left(\frac{\partial V}{\partial T}\right)_P = -\left(\frac{\partial S}{\partial P}\right)_T = -\left(\frac{\partial S}{\partial V}\right)_T \left(\frac{\partial V}{\partial P}\right)_T = -\left(\frac{\partial P}{\partial T}\right)_V \left(\frac{\partial V}{\partial P}\right)_P$$

$$\therefore \quad C_P - C_V = -T\left[\left(\frac{\partial P}{\partial T}\right)_V\right]^2 \left(\frac{\partial V}{\partial P}\right)_T$$

2 Fundamentals of Quantum Statistical Mechanics

QUANTAL MODEL OF MATTER AND STATISTICAL OPERATOR

At very low temperatures, the classical statistical mechanics, i.e., the statistical theory based on classical model of matter, leads to erroneous results. The situation is remedied if we pass over to a statistical theory based on a quantal model of matter. The quantal model means a system of material particles (may be atoms, molecules or elementary particles), whose motion conforms to the laws of quantum mechanics. In classical mechanics, the state of a particle is defined by exact specification of p and q. In quantum mechanics, simultaneous determination of p and q is not possible (uncertainly principle). The state of a particle is defined by the wave function and to every physical quantity (observable), there corresponds a linear operator. The laws of motion in quantum mechanics then enable us to calculate:

(a) the possible values of physical quantities (discrete or continuous) called the eigenvalue spectrum,
(b) the probabilities for specified values of physical quantities and average values, and
(c) the change of probabilities or mean values with time.

The coordinate and momentum operators are

$$\hat{q}_k = q_k, \hat{p}_k = -i\hbar \frac{\partial}{\partial q_k}$$

The operator corresponding to any function of p and q is then found by the above substitution. The possible values (spectrum) of the quantity $\hat{L}$ is given by the eigenvalue equation

$$\hat{L}\psi(q) = L\psi(q) \tag{2.1}$$

where ψ is the wave function representing the state and L is the eigenvalue.

The mean values are given by

$$\langle \hat{L} \rangle = \int \psi^*(q)\hat{L}\psi(q)\,dq \tag{2.2}$$

where ψ^* is the complex conjugate of ψ.

The probability density is

$$P(q) = \psi^*(q)\,\psi(q) \tag{2.3}$$

The time development is given by the Schrödinger equation

$$i\hbar\frac{\partial \Psi(q,t)}{\partial t} = \hat{H}\Psi(q,t) \tag{2.4}$$

where H, the Hamiltonian operator for N-particle system is given by

$$\hat{H} = \sum_{n=1}^{3N}\frac{\hat{p}_n^2}{2m} + U(\hat{q}_1, \ldots, \hat{q}_{3N}) \tag{2.5}$$

The time and space part of the Schrödinger equation can be separated to yield

$$\Psi_n(q,t) = \psi_n(q)e^{-iE_nt/\hbar}$$

$$\hat{H}_n\psi_n(q) = E_n\psi_n(q) \tag{2.6}$$

The basic difference between the quantal picture and classical picture is that quantum mechanics itself is a statistical theory and therefore, quantum statistics investigating a system with many degrees of freedom will envisage a statistical averaging superimposed on the inherent quantum mechanical averaging.

In quantum mechanics, the description of state by means of wave function is most complete and exhaustive. If the macroscopic conditions of a system are such that they entirely determine the state of microparticles, then the state of these particles may be described by a single wave function, and the system is said to be in a pure state. However, in reality, one often meets with the situation in which, from the start, the particles of the system are in different states, described by different wave functions $\Psi_1, \Psi_2, \ldots, \Psi_N$ with probabilities $P_1, P_2, \ldots, P_N$. The state of such a system is called *mixed state* and is described by statistical matrix or density matrix introduced by Neumann. This description is based on incomplete information about the system and the knowledge of statistical matrix permits a calculation of the mean value of any arbitrary quantity of the system as well as the probabilities of the different values of this quantity.

The average value of an arbitrary quantity L is given by

$$\langle L \rangle = \sum_\alpha P_\alpha \langle L_\alpha \rangle = \sum_\alpha P_\alpha \int \Psi_\alpha^* \hat{L}\Psi_\alpha\,dq$$

Using the expansion $\Psi_\alpha = \sum_n c_{an}\psi_n$, we get

$$\langle L \rangle = \sum_\alpha\sum_m\sum_n P_\alpha c_{\alpha m}^* c_{an} \int \psi_m^* \hat{L}\psi_n\,dq \tag{2.7}$$

which can be written as

$$\langle L \rangle = \sum_m \sum_n w_{nm} L_{mn} \tag{2.8}$$

where

$$L_{mn} = \int \psi_m^* \hat{L} \psi_n \, dq; \quad w_{nm} = \sum_\alpha P_\alpha c_{\alpha m}^* c_{\alpha n} = \frac{1}{N} \sum_{\alpha=1}^{N} c_{\alpha m}^* c_{\alpha n} \tag{2.9}$$

The set of quantities w_{nm}, which in general are functions of time, is the *density matrix* in the energy representation or is called *statistical matrix*. If the matrix elements w_{nm} correspond to an operator $\hat{w}$, called statistical operator, then the sum on the right-hand side of Eq. (2.7) will be the diagonal elements of the matrix of the operator $\hat{w}\hat{L}$, i.e. the mean value of $\langle L \rangle$ will be given by the trace of the operator $\hat{w}\hat{L}$:

$$\langle L \rangle = \sum_{m,n} w_{nm} L_{mn} = \sum_n (\hat{w}\hat{L})_{nn} = Tr(\hat{w}\hat{L})$$

$$= \sum_n \int \psi_n(q) \hat{w}\hat{L} \, \psi_n(q) \, dq \tag{2.10}$$

Since the trace of an operator is independent of the choice of the system of functions which defines the matrix elements, the above has the advantage that it allows one to work with any arbitrary complete orthonormal set of wave functions. If we replace L_{mn} by δ_{mn}, the Kroneckar delta

$$\langle L \rangle = 1 = \sum_{m,n} w_{nm} \delta_{mn} = \sum_n w_{nn} = Tr(\hat{w})$$

i.e.,

$$Tr(\hat{w}) = 1 \tag{2.11}$$

Since $w_n = w_{nn}$ is the probability that the system is in nth state, it is always positive, i.e.

$$w_n > 0$$

If we compare Eq. (2.8) with Eq. (1.5), we find that integration over p and q in the classical case has been replaced by the double sum over quantum states and Eq. (2.11) corresponds to normalization condition given by Eq. (1.4). We further show that $\hat{w}$ satisfies a quantum mechanical version of Liouville theorem.

For an arbitrary operator $\hat{L}$, we have

$$\frac{d}{dt}\langle L \rangle = \frac{d}{dt} \int \Psi^* \hat{L} \Psi \, dq$$

$$= \int \Psi^* \frac{\partial \hat{L}}{\partial t} \Psi \, dq + \frac{1}{i\hbar} \int \Psi^* (\hat{L}\hat{H} - \hat{H}\hat{L}) \Psi \, dq$$

(We have used Schrödinger equation (2.4) in the second part.)

This can be written as

$$\frac{d}{dt}\langle L\rangle = \left\langle \frac{\partial L}{\partial t}\right\rangle + \frac{1}{i\hbar}\left\langle \left[\hat{L}, \hat{H}\right]\right\rangle$$

i.e.

$$\left\langle \frac{\partial L}{\partial t}\right\rangle = \left\langle \frac{\partial L}{\partial t}\right\rangle + \left\langle \left\{\hat{L}, \hat{H}\right\}\right\rangle \tag{2.12}$$

where we use quantum Poisson bracket, $\left\{\hat{L}, \hat{H}\right\} = \frac{1}{i\hbar}\left[\hat{L}, \hat{H}\right]$.

We write in operator form

$$\frac{d\hat{L}}{dt} = \frac{\partial \hat{L}}{\partial t} + \left\{\hat{L}, \hat{H}\right\} \tag{2.13}$$

Replacing $\hat{L}$ by $\hat{w}$, we have

$$\frac{d\hat{w}}{dt} = \frac{\partial \hat{w}}{\partial t} + \left\{\hat{w}, \hat{H}\right\} \tag{2.14}$$

In a closed system, the wave function will always have the same form, i.e., will be independent of time, but c will depend on time. We write

$$c_{\alpha n}(t) = c_{\alpha n}(0)\, e^{-iE_n t/\hbar}$$

This will give

$$\begin{aligned}\frac{\partial w_{nm}}{\partial t} &= \frac{i}{\hbar}(E_m - E_n) w_{nm} \\ &= -\frac{1}{i\hbar}\sum_l (w_{nl}H_{lm} - H_{nl}w_{lm})\end{aligned}$$

In operator form, we have

$$\frac{\partial \hat{w}}{\partial t} = -\frac{1}{i\hbar}(\hat{w}\hat{H} - \hat{H}\hat{w}) = -\frac{1}{i\hbar}\left[\hat{w}, \hat{H}\right] = -\left\{\hat{w}, \hat{H}\right\}$$

Alternatively,

$$\frac{\partial \hat{w}}{\partial t} = -i\hat{L}\hat{w} \tag{2.15}$$

where $\hat{L}$ is Liouville operator defined by

$$\hat{L} = -i\sum_k \left(\frac{\partial H}{\partial p_k}\frac{\partial}{\partial q_k} - \frac{\partial H}{\partial q_k}\frac{\partial}{\partial p_k}\right) \equiv -i\sum_k \left(\frac{p_k}{m}\frac{\partial}{\partial q_k} + F_k\frac{\partial}{\partial p_k}\right)$$

From Eqs. (2.14) and (2.15), we find

$$\frac{d\hat{w}}{dt} = 0 \tag{2.16}$$

a result analogous to the classical expression given by Eq. (1.13). These considerations lead us to establish that the statistical matrix or density matrix plays the same role in quantum statistics as the distribution function does in classical statistics.

POSTULATES OF QUANTUM STATISTICAL MECHANICS

We have seen that the operator $\hat{w}$ is a constant of motion; it must be some function of $\hat{H}$, i.e.

$$\hat{w} = \phi(\hat{H}) \tag{2.17}$$

and

$$w_{nm} = \int \psi_n^* \phi(\hat{H})\, \psi_m \, dq \tag{2.18}$$

If we consider $\phi(\hat{H})$ as a power series in $(\hat{H})$,

$$\phi(\hat{H}) = f(E_m) \tag{2.19}$$

and

$$w_{nm} = \delta_{nm} f(E_m) \tag{2.20}$$

Thus, any $f(\hat{H})$ is a diagonal matrix giving the probability of observing the eigenvalue E_m

$$f(E_m) = \langle P_m \rangle \tag{2.21}$$

$$\begin{aligned} \langle P_m \rangle &= 1/N \text{ for the } N \text{ basic states,} \\ &\quad \text{with } E \le E_m < E + \delta E \text{ of a closed isolated system} \\ &= 0 \text{ otherwise} \end{aligned} \tag{2.22}$$

The factor $1/N$ results from the normalization condition. This is usually called the *postulate of equal a priori probabilities.*

In quantum mechanics, ψ is not a measurable quantity but $|\psi|^2$ is. Thus, the wave functions are determined but for a phase factor like e^{iv}. Hence w_{nm} will contain a factor $\langle e^{i(v_m - v_n)} \rangle$. The postulate of random phases means that this factor must vanish for $n \neq m$, i.e.

$$w_{nm} = 0 \quad \text{for } n \neq m \tag{2.23}$$

Thus, the statistical matrix will not be a diagonal matrix if the above condition is not satisfied. In the absence of any information concerning the phases v_m, it is assumed that v_m is assigned random values to make Eq. (2.23) true. This is called the *postulate of random phases.*

MICROCANONICAL DISTRIBUTION IN QUANTUM STATISTICS

We now apply the results obtained so far to a closed system or a quasi-closed system considered only for intervals of time during which they behave with sufficient approximation like a closed system. We have seen that the statistical matrix $w_{nn} = w_n$ must be diagonal. The probabilities w_n now replace the distribution function and the formula for mean value of any quantity, from Eq. (2.8), L becomes simply

$$\langle L\rangle = \sum_n w_n \, L_{nn} \tag{2.24}$$

We proceed in a manner similar to that used in the derivation of Eq. (1.15) for the distribution of subsystem a to obtain

$$\ln w_n^{(a)} = \alpha^{(a)} + \beta E_n^{(a)} \tag{2.25}$$

The number of quantum states of a macroscopic body depends on energy, volume and the number of particles. Since N and V are large, the energy levels will be so close together as to form an almost continuous spectra. We denote by $d\Gamma$ the number of quantum states that belong to a particular infinitesimal dE of its energy. $d\Gamma$ plays a role analogous to the phase volume $dp\ dq$ of the classical case. If we neglect the interaction between the subsystems, then for the whole closed system we can write

$$d\Gamma = \prod_a d\Gamma_a \tag{2.26}$$

The microcanonical distribution is obtained now in a similar manner as in the classical case for the probability of finding the system in any of $d\Gamma$ states as

$$dw = \text{constant}\ \delta(E - E_0) \prod_a d\Gamma_a \tag{2.27}$$

No Gibbs paradox will result in this case, as correct counting of states is already implied in the definition of w_n.

No new result follows from quantal microcanonical distribution that is not obtained from the classical counterpart.

SYSTEM OF IDENTICAL PARTICLES AND SYMMETRY OF WAVE FUNCTION

By identical particles we mean particles that have the same mass, charge, spin, etc. so that they behave in the same manner under similar conditions. In classical mechanics, precise measurement of both p and q of every particle is possible and hence we can construct the path of motion of every particle at least in principle. Hence the particles are distinguishable and can be numbered. In quantum mechanics, the path of motion of the particles cannot be constructed as the simultaneous determination of p and q is limited by uncertainty principle

$$\delta p\ \delta q \geq h \tag{2.28}$$

Hence the particles cannot be numbered and thus become indistinguishable. To get an idea of the classical limit in a system of N identical particles in a volume V where $\langle r\rangle$ and $\langle p\rangle$ denote the average interparticle separation and average particle momentum, from Eq. (2.28) we write

$$\langle p\rangle\langle r\rangle \geq h \tag{2.29}$$

We have

$$\langle r \rangle \sim \left(\frac{V}{N}\right)^{1/3}$$

and

$$\left\langle \frac{p^2}{2m} \right\rangle = \frac{1}{2} kT$$

We get

$$\left(\frac{V}{N}\right)^{1/3} \geq \frac{h}{(mkT)^{1/2}} \tag{2.30}$$

We thus conclude that for classical description V should by large and N small, i.e. a dilute gas, temperature T should be large and mass of the particles not too small. Under such conditions, the interparticle separation will be large and the wave function of the individual particles will not overlap, and it will be possible to distinguish the particles.

Let us represent the coordinates of kth particle by q_k, which includes the position coordinates and spin. The Hamiltonian of the system of N non-interacting identical particles is the sum of Hamiltonions of the single particles, i.e.

$$\hat{H}(q_1, \ldots, q_N, t) = \hat{H}(q_1, t) + \hat{H}(q_2, t) + \cdots + \hat{H}(q_N, t) \tag{2.31}$$

If we define $\hat{P}_{kj}$, a particle interchange operator, that interchanges the kth and jth particles, then

$$\hat{P}_{kj}\hat{H} = \hat{H} \tag{2.32}$$

and

$$\hat{P}_{kj}\,\Psi(q_1, \ldots, q_k, q_j, \ldots, q_n, t) = \Psi(q_1, \ldots, q_j, q_k, \ldots, q_n, t) \tag{2.33}$$

We immediately see that $\hat{P}_{kj}$ and $\hat{H}$ commute and that Ψ and $\hat{P}_{kj}\Psi$ are solutions of the Schrödinger equation, i.e. represent one of the possible states, differing in the distribution of particles among the states. This implies indistinguishability of the particles. The state is defined by a set of the four quantum numbers n, l, m and s. Since Ψ and $\hat{P}_{kj}\Psi$ describe the same physical state, they can differ only by a constant factor.

$$\hat{P}_{kj}\Psi = \lambda\Psi \tag{2.34}$$

Hence

$$\hat{P}_{kj}^2\Psi = \hat{P}_{kj}\hat{P}_{kj}\Psi = \lambda\hat{P}_{kj}\Psi = \lambda^2\Psi = \Psi \tag{2.35}$$

since two interchanges of k and j leaves Ψ unchanged.

This gives

$$\lambda = \pm 1 \tag{2.36}$$

Thus,

$$\hat{P}_{kj}\Psi = \Psi \qquad \text{for } \lambda = +1 \tag{2.37}$$

and

$$\hat{P}_{kj}\Psi = -\Psi \qquad \text{for } \lambda = -1 \tag{2.38}$$

We thus find that the interchange of particles leaves either the wave function unchanged or changes its sign. The wave functions that satisfy Eq. (2.37) are called *symmetric wave functions* and those that satisfy Eq. (2.38) are called *antisymmetric wave functions*. It can be further shown that the symmetric wave function always remains symmetric and the antisymmetric wave function always remains antisymmetric. The division of states in these two classes is absolute and the choice of one class or the other for a given system of particles can be determined only by the nature of the particles. Particles whose spin is an integral multiple of $\hbar$ are described by symmetric wave function, are called *Bose particles* or *bosons*, and obey Bose–Einstein statistics. Particles whose spin are half odd integral multiples of $\hbar$ are described by antisymmetric wave function, are called *Fermi particles* or *fermions*, and obey Fermi-Dirac statistics. The *fermions* obey Pauli exclusion principle.

If we represent by n_1, n_2, ..., etc. the set of four quantum numbers, which characterizes the states of different particles, then for a system of non-interacting (or very weakly interacting) particles, we can expand the wave function as a product of single particle wave functions (the simplest approximation):

$$\Psi_{n_1,\ldots,n_N}(q_1,\ldots,q_N) = \Psi_{n_1}(q_1)\Psi_{n_2}(q_2)\ldots\Psi_{n_N}(q_N) \tag{2.39}$$

This product type wave function needs to be properly symmetrized (because of indistinguishability) for interchange of particles. The symmetrized wave function for the symmetric and antisymmetric cases are

$$\Psi_S = \frac{1}{\sqrt{N!}}\sum_P \lambda^P \Psi_{n_1}(q_1)\Psi_{n_2}(q_2)\ldots\Psi_{n_N}(q_N) \tag{2.40}$$

$$\Psi_A = \frac{1}{\sqrt{N!}}\sum_P \lambda^P \Psi_{n_1}(q_1)\Psi_{n_2}(q_2)\ldots\Psi_{n_N}(q_N)$$

$$\equiv \frac{1}{\sqrt{N!}}\begin{vmatrix} \Psi_{n_1}(q_1) & \Psi_{n_1}(q_2) & \cdots & \Psi_{n_1}(q_N) \\ \Psi_{n_2}(q_1) & \Psi_{n_2}(q_2) & \cdots & \Psi_{n_2}(q_N) \\ \cdots & \cdots & \cdots & \cdots \\ \Psi_{n_N}(q_1) & \Psi_{n_N}(q_2) & \cdots & \Psi_{n_N}(q_N) \end{vmatrix} \tag{2.41}$$

where subscripts S and A stand for symmetric and antisymmetric and the summation is carried over all permutation of the states n_1, n_2, ..., n_N over the particles 1, 2, ..., N, and λ is +1 or –1 depending on whether we consider a system of bosons or fermions and P is the number of pair interchanges leading to the permutation.

MAXWELL–BOLTZMANN, FERMI–DIRAC AND BOSE–EINSTEIN DISTRIBUTIONS

It has already been pointed out that for a macroscopic body, the energy levels (states) are so close that they form a continuum. We divide all the energy states in groups of neighbouring energies. Let g_i be the number of states in the ith group and n_i be the number of particles in these states, all of almost equal energy around ε_i. We calculate the statistical weight $\Delta\Gamma$ of the given macroscopic state, i.e., the number of microscopic ways in which the macroscopic state can be achieved. The entropy is then evaluated and is maximized to correspond to a state of equilibrium of a closed system. If $\Delta\Gamma_i$ is the statistical weight of ith group, this is simply the number of ways in which n_i particles in the group can occupy the g_i states available in the group. The statistical weight of the macroscopic body is

$$\Delta\Gamma = \prod_i \Delta\Gamma_i \tag{2.42}$$

Maxwell–Boltzmann Statistics

The particles are distinguishable and the gas is dilute. Thus the mean occupation number $\langle n \rangle$ of states is small compared with unity, i.e., $n_i \ll g_i$, though any number of particles can occupy any state. The smallness of mean occupation number allows us to assume that all particles are distributed over the different states independent of each other. Of the n_i particles in the ith group, the first one can occupy any of these states, so there are g_i ways. The second and all the subsequent ones can again occupy these states in g_i ways. Hence the total number of ways in which n_i particles can be distributed at random among the g_i states is $g_i\ g_i\ g_i\ \ldots$ (n_i times), i.e. $g_i^{n_i}$ ways. But this includes some identical ways differing only by a permutation of the n_i identical particles whose number is $n_i!$. Thus the statistical weight is

$$\Delta\Gamma_i = \frac{g_i^{n_i}}{n_i!} \tag{2.43}$$

This includes a division by $N!$, called *correct Boltzmann counting*, which results from $N!$ permutation of N indistinguishable, identical particles amongst themselves.

The entropy is thus given by

$$S = k \ln \Delta\Gamma = k \sum_i \ln \Delta\Gamma_i \tag{2.44}$$

This gives

$$S = k \sum_i (n_i \ln g_i - \ln n_i!)$$

We use Stirling formula

$$\ln n! = n \ln n - n \qquad \text{for large } n$$

to get

$$S = k \sum_i (n_i \ln g_i - n_i \ln n_i + n_i) \tag{2.45}$$

In the equilibrium state, the entropy must be maximum, i.e.

$$dS = k\sum_i (\ln g_i - \ln n_i)\, dn_i = 0 \tag{2.46}$$

For an isolated closed system, E, N and V are constants,

$$E = \sum_i \varepsilon_i n_i = \text{constant}, \quad N = \sum_i n_i = \text{constant}$$

Hence

$$dN = \sum_i dn_i = 0 \tag{2.47}$$

and

$$dE = \sum_i \varepsilon_i dn_i = 0 \tag{2.48}$$

Using Lagrange method of unknown multipliers, we have

$$\frac{dS}{k} + \alpha\, dN - \beta\, dE = 0 \tag{2.49}$$

Use of Eq. (2.46) yields

$$f_i = \frac{n_i}{g_i} = e^{\alpha - \beta \varepsilon_i} \tag{2.50}$$

Comparison of Eq. (2.49) with Eq. (1.62) and recalling that V is constant gives

$$\alpha = \frac{\mu}{kT}, \quad \beta = \frac{1}{kT} \tag{2.51}$$

The choice of ith group is arbitrary, Eq. (2.50) holds for any group and we can drop the subscript i to write

$$f = e^{(\mu - \varepsilon)/kT} \tag{2.52}$$

which is the *Maxwell–Boltzmann distribution*.

Fermi–Dirac Statistics

In a Fermi gas, the particles are indistinguishable, have spin as half odd integral multiples of $\hbar$ and are represented by antisymmetric wave function. Each quantum state can contain no more than one particle on account of Pauli exclusion principle. Hence occupation number is 0 or 1 and $n_i \le g_i$. The number of possible ways of distributing n_i indistinguishable particles over g_i states with not more than one particle in each state is simply the number of ways of choosing n_i out of g_i states, i.e. equal to ${}^{g_i}C_{n_i}$. Hence we have

$$\Delta\Gamma_i = {}^{g_i}C_{n_i} = \frac{g_i!}{n_i!(g_i - n_i)!} \tag{2.53}$$

This gives for entropy

$$S = k \ln \Delta\Gamma = k \sum_i \ln \Delta\Gamma_i = k \sum_i \left[\ln g_i! - \ln n_i! - \ln (g_i - n_i)!\right] \tag{2.54}$$

Use of Stirling formula simplifies Eq. (2.54) to

$$S = k \sum_i \left[g_i \ln g_i - n_i \ln n_i - (g_i - n_i) \ln (g_i - n_i) \right] \tag{2.55}$$

This gives

$$\frac{dS}{k} = \sum_i \left[\ln (g_i - n_i) - \ln n_i \right] dn_i \tag{2.56}$$

For equilibrium distribution, we maximize entropy, i.e., we put

$$dS = 0$$

and recall that N and E are constants, i.e.,

$$dN = \sum_i dn_i = 0$$

and

$$dE = \sum_i \varepsilon_i dn_i = 0$$

Using Lagrange method of unknown multipliers, we have

$$\frac{dS}{k} + \alpha\, dN - \beta\, dE = 0 \tag{2.57}$$

Substitution from Eq. (2.56) yields

$$\ln \frac{g_i - n_i}{n_i} = \beta\varepsilon_i - \alpha \tag{2.58}$$

This gives

$$\frac{n_i}{g_i} = \frac{1}{e^{\beta\varepsilon_i - \alpha} + 1} \tag{2.59}$$

This holds for any arbitrary group. Dropping the subscript i and substituting for α and β from Eq. (2.51), we get

$$f = \frac{1}{e^{(\varepsilon - \mu)/kT} + 1} \tag{2.60}$$

The above is *Fermi–Dirac distribution.*

Bose–Einstein Statistics

In a Bose gas, the particles are indistinguishable, have spin as integral multiples of $\hbar$ and are represented by symmetric wave function. Any number of particles can occupy a state, i.e., occupation number is 0, 1, 2, ..., and n_i may be of the same order as g_i or may be even greater. The total number of ways of distributing n_i particles amongst the g_i states is found as follows. The g_i states will have $(g_i - 1)$ partitions. Regarding $(g_i - 1)$ partitions and n_i particles as a total of $(g_i + n_i - 1)$ objects, they can be arranged in $(g_i + n_i - 1)!$ ways amongst themselves. But the n_i particles can be arranged in $n_i!$ ways amongst themselves and $(g_i - 1)$ partitions in $(g_i - 1)!$ ways amongst themselves. Hence the total number of distinguishable arrangements is

$$\frac{(g_i + n_i - 1)!}{n_i!(g_i - 1)!}$$

Hence

$$\Delta\Gamma_i = \frac{(g_i + n_i - 1)!}{n_i!(g_i - 1)!} \tag{2.61}$$

This gives for entropy

$$S = k \ln \Delta\Gamma = k\sum_i \ln \Delta\Gamma_i = k\sum_i \left[\ln\left(g_i + n_i - 1\right)! - \ln n_i! - \ln\left(g_i - 1\right)!\right] \tag{2.62}$$

Use of Stirling formula and that n_i and g_i are very large as compared to unity. Thus, Eq. (2.62) simplifies to

$$S = k\sum_i \left[(g_i + n_i)\ln(g_i + n_i) - n_i \ln n_i - g_i \ln g_i\right] \tag{2.63}$$

This gives

$$\frac{dS}{k} = \sum_i \left[\ln(g_i + n_i) - \ln n_i\right] dn_i \tag{2.64}$$

For equilibrium distribution, we maximize entropy, i.e. put

$$dS = 0$$

and recall that N and E are constants, i.e.

$$dN = \sum_i dn_i = 0$$

$$dE = \sum_i \varepsilon_i dn_i = 0$$

Using Lagrange method of unknown multiplies, we have

$$\frac{dS}{k} + \alpha\, dN - \beta\, dE = 0 \tag{2.65}$$

Substitution from Eq. (2.64) gives

$$\ln \frac{g_i + n_i}{n_i} = \beta\varepsilon_i - \alpha \tag{2.66}$$

This yields

$$\frac{n_i}{g_i} = \frac{1}{e^{\beta \varepsilon_i - \alpha} - 1} \tag{2.67}$$

This holds for any arbitrary group, so we drop the subscript i and substitute from Eq. (2.51) for α and β to obtain

$$f = \frac{1}{e^{(\varepsilon - \mu)/kT} - 1} \tag{2.68}$$

which is *Bose–Einstein distribution.*

We have derived the above results by the method of most probable distributions. The use of Stirling formula in the case of quantum distribution may not be justified, as $(g_i - n_i)$ may not be large. However, we will derive the same results from canonical and grand canonical partition functions in a more rigorous way later.

$\mu < 0$ for bosons and positive or negative for fermions but $|\mu|$ is small for both of them. And hence $e^{(\varepsilon - \mu)/kT} << 1$ for large T. Thus, we can see that at large T

$$\frac{1}{e^{(\varepsilon - \mu)/kT} \pm 1} = \frac{e^{(\mu - \varepsilon a)/kT}}{1 \pm e^{(\mu - \varepsilon)/kT}} \simeq e^{(\mu - \varepsilon)/kT}$$

and both Bose-Einstein distribution and Fermi–Dirac distribution approach Maxwell–Boltzmann distribution.

EXERCISES

1. What is the basic difference between the classical and quantum statistics?
2. What are the postulates of quantum statistical mechanics? Derive the micro canonical distribution in quantum statistics.
3. Discuss the difference between Maxwell–Boltzmann, Fermi–Dirac and Bose–Einstein distributions. Show that in the limiting case, both Fermi–Dirac and Bose–Einstein statistics reduce to Maxwell–Boltzmann statistics.
4. What is statistical matrix? Show that in operator form it satisfies the Liouville equation.
5. What are conditions under which classical statistics would not suffice? What are symmetric and antisymmetric wave functions and how do they lead to two different quantum statistics?

3 Gibbs Canonical Distribution

CLASSICAL CANONICAL DISTRIBUTION

Let us consider an isothermic closed system. Let the system consist of two subsystems, one smaller part but still macroscopic (our subsystem of interest, depicted by subscript s) and the other remaining, a larger part, called the *reservoir* (depicted by subscript r). The wall of separation is thermally conducting but impervious to mass transfer, i.e. an exchange of energy is possible but the number of particles remains constant, thus making N, V and T constant. We apply the microcanonical distribution to the system to write the probability dw that the closed system has a given value of Energy E_0, the subsystem of interest has energy E_s and the reservoir has energy E_r

$$dw = \text{constant}\ \delta(E_s + E_r - E_0)\, d\Gamma_s d\Gamma_r \tag{3.1}$$

where $d\Gamma_s$ and $d\Gamma_r$ are statistical weights of the subsystem of interest and the reservoir.

Our aim is to find the probability density that the closed system is in a state in which the subsystem of interest is in some particular state with energy $E = E(p, q) = H(p, q)$. The required probability density can be found by replacing $d\Gamma_s$ by unity (only one state around E per unit volume of the phase space) in Eq. (3.1), putting $E_s = E$ and integrating over $d\Gamma_r$

$$f(p, q) = \text{constant} \int \delta(E + E_r - E_0)\, d\Gamma_r \tag{3.2}$$

Using

$$d\Gamma_r = \frac{\Delta\Gamma_r}{\Delta E_r}\, dE_r = \frac{e^{S_r/k}}{\Delta E_r}\, dE_r$$

(where $\Delta\Gamma_r$ is the statistical weight corresponding to energy range ΔE_r) in Eq. (3.2) and carrying out the integration, we have

$$f(p, q) = \text{constant}\left(\frac{e^{S_r/k}}{\Delta E_r}\right)_{E_r = E_0 - E} \tag{3.3}$$

The subsystem of interest is small in comparison to the whole closed system, i.e. $E_0 \gg E$, and that ΔE_r changes relatively little with small changes in E_r. Hence $(\Delta E_r)_{E_r = E_0 - E}$ becomes almost a constant, independent of E and is absorbed in the constant. Treating entropy as a function of energy, we make use of Taylor expansion and retain only the linear term, to write

$$S_r(E_0 - E) = S_r(E_0) - E\frac{dS_r(E_0)}{dE_0} \tag{3.4}$$

From Eq. (1.38), this reduces to

$$S_r(E_0 - E) = S_r(E_0) - \frac{E}{T} \tag{3.5}$$

The factor $e^{S_r(E_0)/k}$ is constant and again can be absorbed in the constant to yield

$$f(p, q) = \frac{\text{constant}}{h^{3N}} e^{-E(p, q)/kT} = \frac{\text{constant}}{h^{3N}} e^{-H(p, q)/kT} \tag{3.6}$$

This is the classical canonical distribution first obtained by Gibbs in 1901.

The constant is obtained from the normalization condition

$$\int f(p, q)\, dp\, dq = 1 \tag{3.7}$$

PARTITION FUNCTION AND THERMODYNAMIC FUNCTIONS

The reciprocal of the constant in Eq. (3.6) is called the canonical partition function and is obtained from the normalization condition. For a classical system of N-identical particles at temperature T, the canonical partition function Q_N is given by

$$Q_N = \frac{1}{h^{3N} N!} \int \ldots \int e^{-H(p,q)/kT} \prod_{i=1}^{3N} dp_i \prod_{i=1}^{3N} dq_i \tag{3.8}$$

the degrees of freedom being $3N$. It is obvious that Q_N is function of N, V and T.

The factor h^{3N} is the quantum constant and is included for the reason that the classical distribution could be obtained as a limiting case of quantum distribution. The factor $N!$ is from the correct Boltzmann counting and is a result of indistinguishability of particles.

The classical Hamiltonian can always be written as a sum of the kinetic energy $\sum_i p_i^2/2m$ and the potential energy $U(q_1, \ldots, q_{3N})$, depending on the configuration of all the particles

$$Q_N = \frac{1}{h^{3N} N!} \int \ldots \int e^{-\left(\sum_i p_i^2/2mkT + U/kT\right)} \prod_{i=1}^{3N} dp_i \prod_{i=1}^{3N} dq_i$$

The integration over momenta* can be carried out to yield

$$Q_N = \frac{(2\pi mkT)^{3N/2}}{h^{3N} N!} \int \ldots \int e^{-U/kT} \prod_{i=1}^{3N} dq_i$$

$$= \frac{Z_N}{N!\lambda^{3N}} \tag{3.9}$$

where

$$Z_N = \int_V \ldots \int e^{-U/kT} \prod_{i=1}^{3N} dq_i \tag{3.10}$$

and

$$\lambda = \frac{h}{\sqrt{2\pi mkT}} \tag{3.11}$$

The quantity Z_N is called the configurational partition function and λ is the mean de Broglie wavelength at temperature T.

We rewrite Eq. (3.6) as

$$f(p, q) = \frac{1}{h^{3N} Q_N} e^{-E(p,q)/kT}$$

---

*For reference we give here the values of integrals of the form:

$$I_n = \int_0^\infty dx\, x^n e^{-\alpha x^2}$$

Puting $\alpha x^2 = t$,

$$I_n = \frac{1}{2} \alpha^{-(n+1)/2} \int_0^\infty dt\, t^{(n-1)/2} e^{-t} = \frac{1}{2} \alpha^{-(n+1)/2} \Gamma\left(\frac{n+1}{2}\right)$$

where the Gamma function $\Gamma(k)$ is defined by

$$\Gamma(k) = \int_0^\infty dt\, t^{k-1} e^{-t}$$

$$\Gamma(k+1) = k\Gamma(k) = k! \qquad \text{for } k \text{ a positive integer}$$

and

$$\Gamma\left(\frac{1}{2}\right) = \sqrt{\pi}$$

For $n = 2r$, i.e., an even positive integer

$$I_{2r} = \frac{(2r-1)!!}{2^{r+1}} \sqrt{\frac{\pi}{\alpha^{2r+1}}}$$

where $(2r - 1)!! \equiv 1.3.5.... (2r - 1)$.

For $n = 2r + 1$, i.e., an odd positive integer

$$I_{2r+1} = \frac{r!}{2\alpha^{r+1}}$$

The same integral from $-\infty$ to ∞ is zero if $n = 2r + 1$ and twice the value of the integral from 0 to ∞ if $n = 2r$, including $r = 0$.

to yield

$$E = \langle E \rangle = -kT\,\langle \ln h^{3N} f \rangle - kT \ln Q_N$$

Use of Eqs. (1.24), (1.37) and (1.53) gives us

$$A = -kT \ln Q_N \tag{3.12}$$

or

$$Q_N = e^{-A/kT} \tag{3.13}$$

This enables us to write canonical distribution as

$$f = h^{-3N}\, e^{(A - E)/kT} \tag{3.14}$$

Equation (3.12) serves as a basis for thermodynamic applications of the Gibbs distribution. Other thermodynamic quantities can be represented in terms of the partition function:

$$P = -\left(\frac{\partial A}{\partial V}\right)_T = kT\left(\frac{\partial \ln Q_N}{\partial V}\right)_T \tag{3.15}$$

$$S = -\left(\frac{\partial A}{\partial T}\right)_V = k \ln Q_N + kT\left(\frac{\partial \ln Q_N}{\partial T}\right)_V \tag{3.16}$$

$$G = A + PV = -kT \ln Q_N + kT\left(\frac{\partial \ln Q_N}{\partial \ln V}\right)_T \tag{3.17}$$

$$\Omega = -PV = -kT\left(\frac{\partial \ln Q_N}{\partial \ln V}\right)_T \tag{3.18}$$

and the energy is

$$E = -\frac{\partial \ln Q_N}{\partial \beta} \tag{3.19}$$

where $\beta = 1/kT$.

APPLICATION TO AN IDEAL GAS

For a perfect one-component monoatomic gas of N particles, each of mass m at temperature T, we have

$$E = \sum_{i=1}^{3N} p_i^2/2m$$

and the partition function is given by

$$Q_N = \frac{1}{h^{3N} N!}\int \ldots \int e^{-\sum_i p_i^2/2m} \prod_{i=1}^{3N} dp_i \prod_{i=1}^{3N} dq_i \tag{3.20}$$

Considering that the generalized coordinates in a group of three correspond to each of the particles, we have

$$\int \ldots \int \prod_{i=1}^{3N} = V^N$$

Also since the kinetic energy of whole of the gas is the sum of the kinetic energies of all the individual particles and that the probability of individual particles are independent of each other and fall into a product of factors, Eq. (3.20) reduces to

$$Q_N = \frac{1}{N!}\left(\frac{V}{h^3}\iiint e^{-\vec{p}^2/2mkT}\, d\vec{p}\right)^N$$

This yields

$$Q_N = \frac{q^N}{N!} \tag{3.21}$$

where

$$q = \frac{V}{h^3}\iiint e^{-\vec{p}^2/2mkT}\, d\vec{p} = V\left(\frac{2\pi mkT}{h^2}\right)^{3/2} \tag{3.22}$$

and is called the *single particle partition function.* Substitution of Eq. (3.22) in Eq. (3.21) gives

$$Q_n = \frac{V^N}{N!}\left(\frac{2\pi mkT}{h^2}\right)^{3N/2} \tag{3.23}$$

The Helmholtz free energy is given by Eq. (3.12)

$$A = -NkT\,\ln\left[\frac{Ve}{N}\left(\frac{2\pi mkT}{h^2}\right)^{3/2}\right] \tag{3.24}$$

Use of Eq. (3.15) gives for pressure

$$P = \frac{NkT}{V} \tag{3.25}$$

which is the *equation of state.*

For entropy, from Eq. (3.16), we get

$$S = Nk\,\ln\frac{V}{N}\left(\frac{2\pi mkT}{h^2}\right)^{3/2} + \frac{5}{2}\,Nk$$

The energy, from Eq. (3.19), we get

$$E = \frac{3}{2}\,NkT \tag{3.26}$$

This gives for the specific heat

$$C_V = \frac{3}{2}\,Nk$$

MAXWELL VELOCITY DISTRIBUTION

Recalling that the energy of an ideal gas is the sum of the kinetic energies of the individual particles and that the probability of individual particles are independent of each other and fall into a product of factors, we can write for the probability that a particle has momentum between $\vec{p}$ and $\vec{p} + d\vec{p}$ as

$$dw_p = \text{constant } e^{-\vec{p}^2/2mkT} \, d\vec{p} \tag{3.27}$$

The constant is found from the normalization condition to be $(2\pi mkT)^{-3/2}$. This gives

$$dw_p = \frac{1}{(2\pi mkT)^{3/2}} e^{-\vec{p}^2/2mkT} \, d\vec{p} \tag{3.28}$$

In Cartesian coordinates, this becomes

$$dw_p = \frac{1}{(2\pi mkT)^{3/2}} e^{-(p_x^2+p_y^2+p_z^2)/2mkT} \, dp_x \, dp_y \, dp_z \tag{3.29}$$

In terms of velocities, this can be written as

$$dw_v = \left(\frac{m}{2\pi kT}\right)^{3/2} e^{-m(v_x^2+v_y^2+v_z^2)/2kT} \, dv_x \, dv_y \, dv_z \tag{3.30}$$

Change to spherical polar coordinates yields

$$dw_v = \left(\frac{m}{2\pi kT}\right)^{3/2} e^{-mv^2/2kT} \, v^2 \sin\theta \, dv \, d\theta \, d\phi \tag{3.31}$$

which is the Maxwell distribution of velocities that gives the probability that a particle has velocity between v and $v + dv$ (Fig. 3.1).

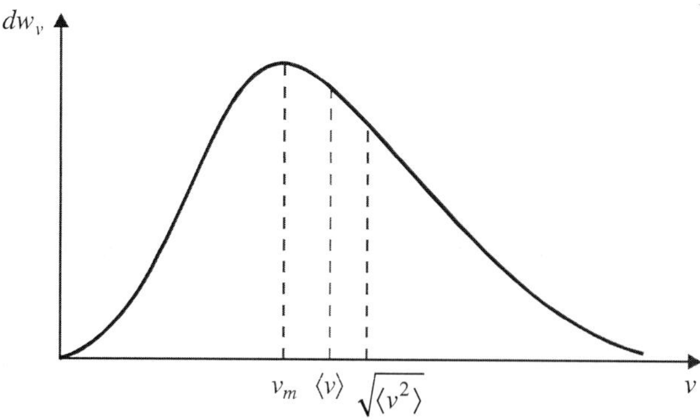

Figure 3.1 Maxwell distribution of velocities.

LAW OF EQUIPARTITION OF ENERGY

The law states that *energy associated with each degree of freedom is* $1/2kT$. Let the energy be written as

$$E = \varepsilon_i + E'(p') \tag{3.32}$$

where ε_i is the energy associated with the momentum coordinate p_i and E' is the energy contributed by all other momenta coordinates except p_i. Then, we have

$$\langle \varepsilon_i \rangle = \frac{\int \varepsilon_i e^{-E/kT} dp}{\int e^{-E/kT} dp} = \frac{\int \varepsilon_i e^{-\varepsilon_i/kT} dp_i}{\int e^{-\varepsilon_i/kT} dp_i} \cdot \frac{\int e^{-E'/kT} dp'}{\int e^{-E'/kT} dp'} \tag{3.33}$$

where

$$dp' = dp_1\, dp_2 \ldots dp_{i-1}\, dp_{i+1} \ldots dp_{3N}$$

The last factor in the above expression is unity, giving us

$$\langle \varepsilon_i \rangle = -\frac{(\partial/\partial\beta)\int e^{-\beta\varepsilon_i} dp_i}{\int e^{-\beta\varepsilon_i} dp_i} = -\frac{\partial}{\partial\beta} \ln\left(\int e^{-\beta\varepsilon_i} dp_i\right) \tag{3.34}$$

where $\beta = 1/kT$. Writing $\varepsilon_i = p_i^2/2m$ and carrying out the integration, we get

$$\langle \varepsilon_i \rangle = -\frac{\partial}{\partial\beta} \ln\left(\frac{2\pi m}{\beta}\right)^{1/2} = \frac{1}{2\beta} = \frac{1}{2} kT \tag{3.35}$$

which is the *law of equipartition of energy*.

QUANTAL CANONICAL DISTRIBUTION

The derivation of classical canonical distribution from microcanonical distribution continues to be valid in quantal case also with the change that integration over Γ space is replaced by a sum over all the states of the system:

$$\frac{1}{h^{3N} N!} \int dp\, dq \rightarrow \sum_n \tag{3.36}$$

Let us consider an isothermic closed system that consists of two subsystems, the smaller but still macroscopic subsystem is the subsystem of interest and is depicted by the subscript s and the larger part, called reservoir, is depicted by the subscript r. The probability that with a given value of energy E_0 of the closed system, the system of interest has energy E_s and the reservoir energy E_r is

$$dw = \text{constant}\ \delta(E_s + E_r - E_0)\, d\Gamma_s\, d\Gamma_r \tag{3.37}$$

where $d\Gamma_s$ and $d\Gamma_r$ refer to the statistical weights of the system of interest and the reservoir respectively.

The probability w_n that the closed system be in a state in which the system of interest is in a particular quantum state with energy E_n is obtained by putting $d\Gamma_s = 1$, $E_s = E_n$ and integrating over $d\Gamma_r$, i.e.,

$$w_n = \text{constant} \int \delta(E_n + E_r - E_0)\, d\Gamma_r \tag{3.38}$$

Using as before

$$d\Gamma_r = \frac{\Delta\Gamma_r}{\Delta E_r}\, dE_r = \frac{e^{S_r/k}}{\Delta E_r}\, dE_r \tag{3.39}$$

The integration yields

$$w_n = \text{constant} \times \left(\frac{e^{S_r/k}}{\Delta E_r}\right)_{E_r = E_0 - E_n} \tag{3.40}$$

Since $E_0 \gg E_n$ and ΔE_r changes relatively little with changes in E_r, $(\Delta E_r)_{E_r = E_0 - E_n}$ is constant and is absorbed in the constant. We make a Taylor expansion of

$$S_r\,(E_r) \equiv S_r(E_0 - E_n)$$

with only the linear term retained:

$$S_r(E_0 - E_n) = S_r(E_0) - E_n \frac{dS_r(E_0)}{dE_0} \tag{3.41}$$

which gives, with aid of Eq. (1.38),

$$S_r(E_0 - E_n) = S_r(E_0) - \frac{E_n}{T} \tag{3.42}$$

The factor $e^{S_r(E_0)/k}$ is constant and can be absorbed in the constant in Eq. (3.40) to give

$$w_n = \text{constant}\ e^{-E_n/kT} = \text{constant} \int \psi_n^*(\vec{r}_1, \ldots, \vec{r}_N)\, e^{-\hat{H}/kT}\, \psi_n(\vec{r}_1, \ldots, \vec{r}_N) \prod_{i=1}^{N} d\vec{r}_i \tag{3.43}$$

$$\equiv \text{constant}\ \langle \psi_n | e^{-\hat{H}/kT} | \psi_n \rangle$$

The constant is found from the normalization condition

$$\sum_n w_n = 1 \tag{3.44}$$

The reciprocal of the constant is called partition function Q_N and is given by

$$Q_N = \sum_n e^{-E_n/kT} = \sum_n \int \psi_n^*\, e^{-\hat{H}/kT}\, \psi_n \prod_{i=1}^{N} d\vec{r}_i$$

$$\equiv \sum_n \langle \psi_n | e^{-\hat{H}/kT} | \psi_n \rangle = Tr(e^{-\hat{H}/kT}) \tag{3.45}$$

The sum on the right-hand side is sum over states and not over energy eigenvalues. The quantum mechanical analogue of configurational partition function is written as

$$Z_N = \int \ldots \int S_N\, d\vec{r} \ldots d\vec{r}_N \tag{3.46}$$

where S_N called *Slater sum* is defined by

$$S_N = N!\,\lambda^{3N} \sum_n \psi^*(\vec{r}_1,\ldots,\vec{r}_N)\, e^{-\hat{H}/kT}\, \psi(\vec{r}_1,\ldots,\vec{r}_N) \tag{3.47}$$

The classical counter part of the Slater sum is the Boltzmann factor $e^{-U/kT}$.

The thermodynamic quantities can be expressed as in the classical case, with w_n replacing f,

$$A = \langle E \rangle - TS = -kT \ln Q_N = -kT \ln\left(\sum_n e^{-E_n/kT}\right) \tag{3.48}$$

$$P = -\left(\frac{\partial A}{\partial V}\right)_T = kT\left(\frac{\partial \ln Q_N}{\partial V}\right)_T \tag{3.49}$$

$$S = -\left(\frac{\partial A}{\partial T}\right)_V = k \ln Q_N + kT\left(\frac{\partial \ln Q_N}{\partial T}\right)_V \tag{3.50}$$

$$G = A + PV = kT\left[-\ln Q_N + \left(\frac{\partial \ln Q_N}{\partial \ln V}\right)_T\right] \tag{3.51}$$

and

$$\Omega = -PV = -kT\left(\frac{\partial \ln Q_N}{\partial \ln V}\right)_T \tag{3.52}$$

GENERIC DISTRIBUTION FUNCTION

For a classical system of one-component monoatomic N particles in a volume V at temperature T, the probability that the particles have momenta $\vec{p}_1,\ldots,\vec{p}_N$ in $d\vec{p}_1,\ldots,d\vec{p}_N$ and position coordinates $\vec{r}_1, \vec{r}_2,\ldots,\vec{r}_N$ in $d\vec{r}_1, d\vec{r}_2,\ldots,d\vec{r}_N$ of a canonical ensemble is given by Gibbs distribution as

$$dw(\vec{p}_1,\ldots,\vec{p}_N,\vec{r}_1,\ldots,\vec{r}_N) = \frac{1}{N!h^{3N}\,Q_N} e^{-H/kT}\, d\vec{p}_1,\ldots,\vec{p}_N, d\vec{r}_1,\ldots, d\vec{r}_N \tag{3.53}$$

with

$H = \sum_{i=1}^{N} \frac{\vec{p}_i^{\,2}}{2m} + U(\vec{r}_1,\ldots,\vec{r}_N)$, the sum of kinetic energy and the potential energy.

The probability of finding the system in a configuration $\vec{r}_1,\ldots,\vec{r}_N$ with one particle in $d\vec{r}_1$ at $\vec{r}_1$ and other particles in $d\vec{r}_2$ at $\vec{r}_2,\ldots,$ is obtained by integrating Eq. (3.53) over the momenta to yield

$$dw\,(\vec{r}_1,\ldots,\vec{r}_N) = \frac{e^{-U(\vec{r}_1,\ldots,\vec{r}_N)/kT}}{Z_N}\, d\vec{r}_1 \,\ldots\, d\vec{r}_N \tag{3.54}$$

The s particle probability density, also called the *generic distribution function*, is defined as the probability density of finding a particle in $d\vec{r}_1$ at $\vec{r}_1$, a second in $d\vec{r}_2$ at $\vec{r}_2,\ldots$, and another in $d\vec{r}_s$ at $\vec{r}_s$, irrespective of the configuration of the remaining $(N - s)$ particles. This can be obtained from Eq. (3.54) as

$$\rho_s(\vec{r}_1,\ldots,\vec{r}_s) = \frac{N!}{(N-s)!} \frac{\int_v \cdots \int e^{-U(\vec{r}_1,\ldots,\vec{r}_N)/kT} \prod_{i=s+1}^{N} d\vec{r}_i}{Z_N} \tag{3.55}$$

The factor $N!/(N - s)!$ arises because any of the N particles can be in $d\vec{r}_1$, any of the remaining $(N - 1)$ in $d\vec{r}_2,\ldots$, and any of the remaining $(N - s + 1)$ in $d\vec{r}_s$, i.e., a total of $N(N - 1) \cdots (N - s + 1)$ ways, i.e. $N!/(N - s)!$ choices.

It is immediately seen that

$$\int_V \cdots \int \rho_s(\vec{r}_1,\ldots,\vec{r}_s)\, d\vec{r}_1 \ldots d\vec{r}_s = \frac{N!}{(N-s)!} \tag{3.56}$$

The simplest distribution function $\rho_1(\vec{r}_1)$ is the probability that one of the particles of the system is in $d\vec{r}_1$ at $\vec{r}_1$. In a fluid, all points $\vec{r}_1$ inside the volume V are equivalent and thus $\rho_1(\vec{r}_1)$ is independent of $\vec{r}_1$, is equal to number density ρ of the fluid, i.e.

$$\frac{1}{V}\int_V \rho_1(\vec{r}_1)\, d\vec{r}_1 = \rho_1 = \frac{N}{V} = \rho \tag{3.57}$$

Extensively used pair distribution function $\rho_2(\vec{r}_1, \vec{r}_2)$ is the probability density of finding one particle in $d\vec{r}_1$ at $\vec{r}_1$ and another particle in $d\vec{r}_2$ at $\vec{r}_2$, irrespective of the configuration of the remaining $(N - 2)$ particles. In a uniform fluid, $\rho_2(\vec{r}_1, \vec{r}_2)$ can depend only on $r_{12} = |\vec{r}_1 - \vec{r}_2|$ and we can write

$$\rho_2(\vec{r}_1, \vec{r}_2) = \rho_2(r_{12}) \tag{3.58}$$

Also,

$$\int_V \int \rho_2(\vec{r}_1, \vec{r}_2)\, d\vec{r}_1\, d\vec{r}_2 = V \int_V \rho_2(\vec{r}_{12})\, d\vec{r}_{12} = N(N-1) \tag{3.59}$$

and for a completely random distribution of particles, we have

$$\rho_1 = \frac{N}{V} = \rho \tag{3.60}$$

and

$$\rho_2 = \frac{N(N-1)}{V^2} = \rho^2\left(1 - \frac{1}{N}\right) \tag{3.61}$$

If the probabilities of finding a particle in $d\vec{r}_1$ at $\vec{r}_1$, a particle in $d\vec{r}_2$ at $\vec{r}_2,\ldots$, were not all independent, we must multiply by a correlation factor. The correlation factor $g_s(\vec{r}_1,\ldots,\vec{r}_s)$ is defined by

$$\rho_s(\vec{r}_1,\ldots,\vec{r}_s) = \rho_1(\vec{r}_1)\ldots\rho_1(\vec{r}_s)\, g_s(\vec{r}_1,\ldots,\vec{r}_s) \tag{3.62}$$

In a fluid, this simplifies to

$$\rho_s(\vec{r}_1,\ldots,\vec{r}_s) = \rho^s g_s(\vec{r}_1,\ldots,\vec{r}_s) \tag{3.63}$$

We further have the pair correlation function (identical with radial distribution function):

$$g_2(\vec{r}_1,\vec{r}_2) = g(\vec{r}_{12}) \tag{3.64}$$

and

$$\rho_2(\vec{r}_{12}) = \rho^2 g(\vec{r}_{12}) \tag{3.65}$$

To get a physical meaning of the pair correlation function, let a particle be fixed at $\vec{r}_1$. Then, the probability of observing a second particle in $d\vec{r}_2$ at $\vec{r}_2$ is

$$\rho g_2(r_{12})\, d\vec{r}_{12} = \rho g(r) 4\pi r^2 dr$$

and we have

$$\rho \int g(r)\, 4\pi r^2 dr = N - 1$$

The thermodynamic functions can be written in terms of the radial distribution function.

$$\frac{E}{NkT} = \frac{3}{2} + \frac{\rho}{2kT}\int_0^\infty u(r)\, g(r)\, d\vec{r} \tag{3.66}$$

and

$$\frac{P}{kT} = \rho - \frac{\rho^2}{6kT}\int_0^\infty r\frac{du(r)}{dr}\, g(r)\, d\vec{r} \tag{3.67}$$

where we have used the approximation

$$U(\vec{r}_1,\ldots,\vec{r}_N) = \sum_{1\le i<j\le N} u(r_{ij}) \tag{3.68}$$

and

$$r_{ij} \equiv r$$

For a quantal system, the s-particle distribution function is obtained as

$$\rho_s(\vec{r}_1,\ldots,\vec{r}_s) = \frac{N!}{(N-s)!}\,\frac{\int\ldots\int S_N \prod_{i=s+1}^{N} d\vec{r}_i}{\mathcal{Z}_N} \tag{3.69}$$

where S_N is Slater sum defined by Eq. (3.47) and $\mathcal{Z}_N$ is quantum mechanical analogue of configurational partition function.

Normalization gives (as in classical case)

$$\int_V \cdots \int \rho_s(\vec{r}_1,\ldots,\vec{r}_s)\, d\vec{r}_1 \ldots d\vec{r}_s = \frac{N!}{(N-s)!} \tag{3.70}$$

Other relations of $\rho_2(r_{12})$ and $g_2(r_{12})$ in the quantal system remain as they are in the classical system.

PARTITION FUNCTION OF DIATOMIC GASES

The energy of diatomic gas has contributions from the translational motion, the rotational motion, the vibrational motion, the electron energy part and the nuclear energy part and thus can be expressed as a sum of all these:

$$E = E_{(t)} + E_{(r)} + E_{(v)} + E_{(e)} + E_{(n)} \tag{3.71}$$

where subscripts t, r, v, e, n stand for translational, rotational, vibrational, electronic and nuclear parts respectively.

The partition function has the product structure:

$$Q_N = Q_t\, Q_r\, Q_v\, Q_e\, Q_n \tag{3.72}$$

and the single particle (a single diatomic molecule) partition function is given by

$$q = q_t\, q_r\, q_v\, q_e\, q_n \tag{3.73}$$

The translational energy levels of a particle enclosed in a cubic box of dimensions L_x, L_y and L_z is given by

$$\varepsilon_n = \frac{\pi^2\hbar^2}{2m}\left(\frac{n_x^2}{L_x^2} + \frac{n_y^2}{L_y^2} + \frac{n_z^2}{L_z^2}\right) \tag{3.74}$$

where n_x, n_y and n_z are quantum numbers having values 1, 2, 3, ...

This gives

$$q_t = \sum_{n_x=1}^{\infty}\sum_{n_y=1}^{\infty}\sum_{n_z=1}^{\infty} e^{-\frac{\pi^2\hbar^2}{2m}\left(\frac{n_x^2}{L_x^2} + \frac{n_y^2}{L_y^2} + \frac{n_z^2}{L_z^2}\right)/kT} \tag{3.75}$$

At ordinary temperatures, the levels are very close to each other and form almost a continuum. We can thus replace the summation by integration that yields

$$q_t = V\left(\frac{2\pi mkT}{h^2}\right)^{3/2} \tag{3.76}$$

where we have used $L_x\, L_y\, L_z = V$.

The result is same as that for a classical monoatomic gas. The free energy is given by Eq. (3.24).

The rotational energy levels are given by

$$\varepsilon_r = \frac{\hbar^2}{2I}\, j(j+1), \quad j = 0, 1, 2, \ldots \tag{3.77}$$

where I is the moment of inertia of the molecule and j is the rotational quantum number.

For a given j, the quantum number m takes $(2j + 1)$ values, giving rise to a $(2j + 1)$ fold degeneracy. Hence we can write for rotational part

$$q_r = \sum_{j=0}^{\infty}(2j+1)\, e^{-(\hbar^2/2I)\, j(j+1)/kT} \tag{3.78}$$

We can again replace the summation by integration, as the energy levels are very closely spaced, to get

$$q_r = \frac{2IkT}{\hbar^2}\sigma^{-1} \tag{3.79}$$

where σ is symmetry number and is equal to 1 for a symmetrical molecule and 2 for asymmetrical molecule.

The free energy is given by

$$A_r = -NkT \ln\left[\frac{2IkT}{\sigma\hbar^2}\right] \tag{3.80}$$

If the temperature is not very large, the vibrations are small and harmonic. The energy levels of a harmonic oscillator is given by

$$\varepsilon_v = \left(\gamma + \frac{1}{2}\right)h\nu, \quad \gamma = 0, 1, 2, \ldots \tag{3.81}$$

If we take the zero point energy as the zero of the scale, then

$$q_v = \sum_{v=0}^{\infty} e^{-\gamma h\nu/kT}$$

This gives

$$q_v = \frac{1}{1 - e^{-h\nu/kT}} \tag{3.82}$$

and the free energy

$$A_v = NkT \ln(1 - e^{-h\nu/kT}) \tag{3.83}$$

If there is thermal equilibrium among electronic states, the electronic partition function is

$$q_e = \sum_k g_k\, e^{-\varepsilon_k/kT} \tag{3.84}$$

where g_k is the number of electronic states in the kth level of energy ε_k. Information about the levels is obtained from spectroscopic experiments. The spacing between electronic energy levels is usually so large that the corresponding characteristic temperature is of the order of thousands of kelvin, so one needs to take into account only few levels.

The situation for the nuclear states is similar to that for the electronic states, except that the characteristic temperature corresponding to the excitation energies of the nuclear states are of the order of billions of kelvin. Hence the nuclear contribution to the one particle partition function is simply the multiplicity of the nuclear ground state of the molecule:

$$q_n = \prod_i g_i \tag{3.85}$$

where g_i is the multiplicity of the ground state of the ith nucleus in the molecule. If in a diatomic molecule, the nuclei have spins s_1 and s_2, then

$$g_i = (2_{s1} + 1)\,(2_{s2} + 1)$$

EXERCISES

1. Derive Gibbs canonical distribution and apply it to a perfect gas to obtain its equation of state.
2. What is the canonical partition function? How are thermodynamic functions related to it?
3. Apply the canonical distribution to an ideal gas to obtain the Maxwell distribution of velocities.
4. Derive equipartition law of energy using the canonical distribution.
5. Discuss the partition function of diatomic molecules.
6. Calculate the average speed of the particles of an ideal gas.

   Hint: $$\langle v\rangle = 2\times 4\pi\left(\frac{m}{2\pi kT}\right)^{3/2}\int_0^\infty e^{-mv^2/kT}v^3\,dv = \sqrt{\frac{8kT}{\pi m}}$$

7. Calculate the average kinetic energy of the particles of an ideal gas.

   Hint: $$\langle v^2\rangle = 4\pi\left(\frac{m}{2\pi kT}\right)^{3/2}\int_0^\infty e^{-mv^2/kT}v^4\,dv = \frac{3kT}{m}$$

   $$\therefore\quad \langle E\rangle = \frac{1}{2}m\,\langle v^2\rangle = \frac{3}{2}kT$$

8. Find the most probable velocity of particles of perfect gas.
   Hint: The most probable velocity is given by

   $$\frac{d}{dv}(v^2e^{-mv^2/2kT}) = 0 \quad \text{for } v = v_m$$

   This gives

   $$v_m = \sqrt{\frac{2kT}{m}}$$

9. Prove

   (a) $\langle E^2\rangle = \dfrac{1}{Q_N}\dfrac{\partial^2 Q_N}{\partial\beta^2}$

   (b) $\langle E^2\rangle - \langle E\rangle^2 = \dfrac{\partial^2 \ln Q_N}{\partial\beta^2}$

   (c) $C_v = -(\langle E^2\rangle - \langle E^2\rangle)\dfrac{\partial\beta}{\partial T}$

   Hint:

   (a) $$\frac{1}{Q_N}\frac{\partial^2 Q_N}{\partial\beta^2} = \frac{1}{Q_N}\frac{\partial}{\partial\beta}\left[\frac{\partial}{\partial\beta}\frac{1}{h^{3N}N!}\int e^{-\beta E}dp\,dq\right] = -\frac{1}{Q_N}\frac{1}{h^{3N}N!}\frac{\partial}{\partial\beta}\int Ee^{-\beta E}dp\,dq$$

   $$= \frac{1}{h^{3N}N!}\frac{1}{Q_N}\int E^2e^{-\beta E}dp\,dq = \langle E^2\rangle$$

(b) $$\frac{\partial^2 \ln Q_N}{\partial \beta^2} = \frac{\partial}{\partial \beta}\left[\frac{\partial}{\partial \beta}\ln Q_N\right] = \frac{1}{Q_N}\frac{\partial^2 Q_N}{\partial \beta^2} - \frac{1}{Q_N^2}\left(\frac{\partial Q_N}{\partial \beta}\right)^2 = \langle E^2\rangle - \langle E\rangle^2$$

(c) $$C_v = \left(\frac{\partial E}{\partial T}\right)_V = \left(\frac{\partial \langle E\rangle}{\partial T}\right)_V = -\frac{\partial}{\partial T}\left(\frac{\partial \ln Q_N}{\partial \beta}\right)$$

$$= -\frac{\partial}{\partial \beta}\left(\frac{\partial \ln Q_N}{\partial \beta}\right)\frac{\partial \beta}{\partial T} = -(\langle E^2\rangle - \langle E\rangle^2)\frac{\partial \beta}{\partial T}$$

10. Prove

(a) $E = \frac{3}{2}NkT + \frac{1}{2}N\rho\int_0^\infty u(r)\, g(r)\, d\vec{r}$

(b) $P = \rho kT\left[1 - \frac{\rho}{6kT}\right]\int_0^\infty r\frac{du(r)}{dr}g(r)\, d\vec{r}$

Hint: (a) $E = kT^2\left(\frac{\partial \ln Q_N}{\partial T}\right)_{N,V}$, $Q_N = \frac{Z_N}{N!\lambda^{3N}}$

This gives

$$E = \frac{3}{2}NkT + kT^2\left(\frac{\partial \ln Z_N}{\partial T}\right)_{N,V} = \frac{3}{2}NkT + \langle U\rangle$$

where $\langle U\rangle = \dfrac{\int_V \cdots \int e^{-U/kT} U\, d\vec{r}_1 \ldots d\vec{r}_N}{Z_N}$, the mean total potential energy.

Writing

$$U = \sum_{1\le i<j\le N} u(r_{ij})$$

a sum of $N(N-1)/2$ terms, all of which will give the same result on integration, we have

$$\langle U\rangle = \frac{N(N-1)}{2}\frac{\int\cdots\int e^{-U/kT}u(r_{12})\, d\vec{r}_1 \ldots d\vec{r}_N}{Z_N} = \frac{1}{2}\iint u(r_{12})\,\rho_2(r_{12})\, d\vec{r}_1\, d\vec{r}_2$$

$$= \frac{1}{2}N\rho\int_0^\infty u(r)\, g(r)\, d\vec{r}$$

We have replaced the upper limit by ∞, since $u(r) \to 0$ rapidly as $r \to \infty$.

(b) $$P = kT\left(\frac{\partial \ln Q_N}{\partial V}\right)_{N,T} = kT\left(\frac{\partial \ln Z_N}{\partial V}\right)_{N,T} = kT\frac{1}{Z_N}\left(\frac{\partial Z_N}{\partial V}\right)_{N,T}$$

In order to differentiate $\mathcal{Z}_N$ with respect to V, we have to change the variables of integration (in the expression for $\mathcal{Z}_N$) in such a way that the limits of integration become constants. Let

$$\left.\begin{aligned} x_k &= V^{1/3}x'_k \\ y_k &= V^{1/3}y'_k \\ z_k &= V^{1/3}z'_k \end{aligned}\right\} \quad \text{where } k = 1, 2, \ldots, N$$

and

$$\begin{aligned} r_{ij} &= [(x_i - x_j)^2 + (y_i - y_j)^2 + (z_i - z_j)^2]^{1/2} \\ &= V^{1/3}[(x'_i - x'_j)^2 + (y'_i - y'_j)^2 + (z'_i - z'_j)^2]^{1/2} \end{aligned}$$

Also,

$$\mathcal{Z}_N = V^N \int_0^1 \cdots \int_0^1 e^{-U/kT} dx'_1 \ldots dz'_N$$

$$\therefore \quad \left(\frac{\partial \mathcal{Z}_N}{\partial V}\right)_{N,T} = NV^{N-1}\int_0^1 \cdots \int_0^1 e^{-U/kT} dx'_1 \ldots dz'_N - \frac{V^N}{kT}\int_0^1 \cdots \int_0^1 e^{-U/kT} \frac{\partial U}{\partial V} dx'_1 \ldots dz'_N$$

$$= NV^{N-1}\int_0^1 \cdots \int_0^1 e^{-U/kT} dx'_1 \ldots dz'_N - \frac{V^N}{kT}\int_0^1 \cdots \int_0^1 e^{-U/kT} \sum_{1 \le i < j \le N} \frac{\partial u(r_{ij})}{\partial r_{ij}} \frac{\partial r_{ij}}{\partial V} dx'_1 \ldots dz'_N$$

$$= NV^{N-1}\int_0^1 \cdots \int_0^1 e^{-U/kT} dx'_1 \ldots dz'_N - \frac{V^N}{kT}\int_0^1 \cdots \int_0^1 e^{-U/kT} \sum \frac{r_{ij}}{3V} \frac{\partial u(r_{ij})}{\partial r_{ij}} dx'_1 \ldots dz'_N$$

In the sum $N(N-1)/2$, equivalent terms like $\dfrac{r_{12}}{3V}\dfrac{\partial u(r_{12})}{\partial r_{12}}$ are present.

We now transform back to original variables to get

$$\left(\frac{\partial \ln \mathcal{Z}_N}{\partial V}\right)_{N,T} = \frac{N}{V} - \frac{1}{6V\,kT}\int_V\int r_{12}\frac{\partial u(r_{12})}{\partial r_{12}} \rho_2(r_{12})\, d\vec{r}_1\, d\vec{r}_2$$

$$\therefore \quad P = \rho kT\left[1 - \frac{\rho}{6kT}\right]\int_0^\infty r\frac{\partial u(r)}{\partial r} g(r)\, d\vec{r}$$

4 Grand Canonical Distribution

CLASSICAL GRAND CANONICAL DISTRIBUTION

In chemical processes, the number of particles changes. Also, in quantum process, the number of particles changes due to creation and annihilation of particles. This necessitates the generalization of Gibbs distribution to bodies with variable number of particles. In deriving canonical distribution from microcanonical distribution, we divided the isothermic closed system in two parts, called subsystem of interest and the reservoir, and allowed the exchange of energy but not of the particles. We now relax the condition and also allow the exchange of particles. Thus, in the grand canonical ensemble, both E and N are variables. However, the chemical potential remains constant, thus making μ, V and T constant. It is also referred to as open system. The distribution function will now be a function of the energy of the state of the subsystem as well as the number of particles of the subsystem. This is achieved by making the energy of subsystem a function of the number of particles. We now apply the microcanonical distribution to the closed system to write the probability dw that the closed system has the given value of energy E_0 and number of particles N_0, the subsystem of interest has energy E_{sN_s} (with N_s particles) and the reservoir has energy E_{rN_r} (with N_r particles):

$$dw = \text{constant } \delta(E_{sN_s} + E_{rN_r} - E_0)\, d\Gamma_s\, d\Gamma_r \tag{4.1}$$

We find the probability density that the closed system is in a state in which the subsystem of interest is in some particular state with energy E_N (with N particles) by putting $d\Gamma_s = 1$, $E_{sN_s} = E_N$ and integrating over $d\Gamma_r$:

$$f(p, q, N) = \text{constant} \int \delta(E_N + E_{rN_r} - E_0)\, d\Gamma_r \tag{4.2}$$

Using

$$d\Gamma_r = \frac{\Delta\Gamma_r}{\Delta E_{rN_r}} dE_{rN_r}$$

and carrying out the integration, we have

$$f(p, q, N) = \text{constant}\left(\frac{e^{S_r/k}}{\Delta E_{rN_r}}\right)_{E_{rN_r}=E_0-E_N} \tag{4.3}$$

We again take ΔE_{rN_r} as relatively constant as before and absorb it in the constant and consider the entropy of the reservoir a function of both—its energy equal to $(E_0 - E_N)$ and its number of particles equal to $(N_o - N)$—and make a Taylor expansion in two variables.

$$S_r(E_0 - E_N, N_0 - N) = S_r(E_0, N_0) - E_N\left(\frac{\partial S_r(E_0)}{\partial E_0}\right)_{V,N_0} - N\left(\frac{\partial S_r(N_0)}{\partial N_0}\right)_{V,E_0} \tag{4.4}$$

We have retained only the linear terms in the expansion.

Substitution from Eq. (1.62) gives

$$S_r(E_0 - E_N, N_0 - N) = S_r(E_0, N_0) - \frac{E_N}{T} - \frac{\mu N}{T} \tag{4.5}$$

The factor $e^{S_r(E_0,N_0)/k}$ is a constant and is absorbed in the constant to obtain

$$f(p,q,N) = \frac{\text{constant}}{h^{3N}} e^{(\mu N - E_N)/kT} \tag{4.6}$$

This is the classical grand canonical distribution.

The constant is obtained from the normalization condition:

$$\sum_{N\geq 0} \int f(p,q,N)\, dp^N\, dq^N = 1 \tag{4.7}$$

where $dp^N\, dq^N$ stands for all dp's and dq's when the system has N particles.

PARTITION FUNCTION AND THERMODYNAMIC FUNCTIONS

The reciprocal of the constant in Eq. (4.6) is called the grand canonical partition function Ξ and is given by

$$\Xi = \sum_{N\geq 0} \frac{e^{\mu N/kT}}{N!\, h^{3N}} \int\cdots\int e^{-E_N(p^N, q^N)/kT}\, dp^N\, dq^N \tag{4.8}$$

In terms of canonical partition function Q_N, we have

$$\Xi = \sum_{N\geq 0} e^{\mu N/kT} Q_N = \sum_{N\geq 0} \frac{z^N Z_N}{N!} \tag{4.9}$$

where z is fugacity given by

$$z = \frac{e^{\mu/kT}}{\lambda^3} = \frac{z'}{\lambda^3} \tag{4.10}$$

where $z' = e^{\mu/kT}$ is called absolute activity and $\mathcal{Z}_N$ is the configurational partition function.

The relation between the chemical potential and the Helmholtz free energy

$$\mu = \left(\frac{\partial A}{\partial N}\right)_{V,T}$$

gives

$$\mu = A_{N+1} - A_N = kT \ln\left(\frac{Q_N}{Q_{N+1}}\right) \tag{4.11}$$

which gives

$$z = \frac{(N+1)!}{N!} \frac{\mathcal{Z}_N}{\mathcal{Z}_{N+1}} = \frac{z'}{\lambda^3} \tag{4.12}$$

Also, we have

$$S = -k\langle \ln h^{3N} f\rangle = -\frac{\mu\langle N\rangle - \langle E_N\rangle}{T} + k \ln \Xi$$

Since $\langle N\rangle = N$ and $\langle E_N\rangle = E_N$, we obtain

$$-kT \ln \Xi = E_N - TS - \mu N = A - \mu N = \Omega \tag{4.13}$$

This gives

$$f(p,q,N) = \hbar^{-3N} e^{(\Omega - \mu N - E_N)/kT} \tag{4.14}$$

Other thermodynamic quantities can be obtained from the grand potential

$$S = -\left(\frac{\partial \Omega}{\partial T}\right)_{\mu,V} = k\left(\frac{\partial (T \ln \Xi)}{\partial T}\right)_{\mu,V} \tag{4.15}$$

$$P = \frac{kT}{V} \ln \Xi \tag{4.16}$$

$$N = -\left(\frac{\partial \Omega}{\partial \mu}\right)_{V,T} = kT\left(\frac{\partial \ln \Xi}{\partial \mu}\right)_{V,T} = z\left(\frac{\partial \ln \Xi}{\partial z}\right)_{V,T} \tag{4.17}$$

$$A = \mu N + \Omega = NkT \ln(\lambda^3 z) - kT \ln \Xi \tag{4.18}$$

$$G = \mu N = NkT \ln (\lambda^3 z) \tag{4.19}$$

QUANTAL GRAND CANONICAL DISTRIBUTION

The generalization of Gibbs distribution to quantal model of bodies with variable number of particles can be done exactly in the same manner as in the classical model by considering a subsystem of interest and the reservoir, both parts of the closed isothermic system. The only difference will be that one has to take the energy quantum states to vary with the number of particles and denote it by E_{nN}, the energy in the nth quantum state. The probability w_{nN} that the subsystem of interest contains N particles and is in the nth state is obtained in the same way to be

$$w_{nN} = \text{constant } e^{S_r(E_0 - E_{nN},\, N_0 - N)/k} \tag{4.20}$$

We again make Taylor expansion in two variables and absorb $e^{S_r(E_0,\, N_0)/k}$ in the constant to get

$$w_{nN} = \text{constant } e^{(\mu N - E_{nN})/k} \tag{4.21}$$

which is the quantal grand canonical distribution. The constant is again found by normalization condition with integration over p, q being replaced by summation over the states to give the quantal grand canonical partition function

$$\Xi = \sum_{N \geq 0} \sum_{n} e^{(\mu N - E_{nN})/kT} = \sum_{N \geq 0} e^{\mu N/kT} Q_N = \sum_{N \geq 0} \frac{z^N \mathcal{Z}_N}{N!} \tag{4.22}$$

where $\mathcal{Z}_N$ is now quantal analogue of the configurational partition function given by Eq. (3.46).

We have

$$S = -k \ln w_{nN} = -\frac{\mu N - E_{nN}}{T} + k \ln \Xi \tag{4.23}$$

This gives

$$-kT \ln \Xi = E_{nN} - TS - \mu N = A - \mu N = \Omega \tag{4.24}$$

Hence we can rewrite Eq. (4.21) as

$$w_{nN} = e^{(\Omega + \mu N - E_{nN})/kT} \tag{4.25}$$

With the aid of Eq. (4.24), we have for other thermodynamic quantities

$$S = -\left(\frac{\partial \Omega}{\partial T}\right)_{\mu, V} = k\left(\frac{\partial (T \ln \Xi)}{\partial T}\right)_{\mu, V} + k \ln \Xi \tag{4.26}$$

$$P = \frac{kT}{V} \ln \Xi \tag{4.27}$$

$$N = -\left(\frac{\partial \Omega}{\partial \mu}\right)_{V,T} = kT\left(\frac{\partial \ln \Xi}{\partial \mu}\right)_{V,T} = z\left(\frac{\partial \ln \Xi}{\partial z}\right)_{V,T} \tag{4.28}$$

$$A = \mu N + \Omega = NkT \ln(\lambda^3 z) - kT \ln \Xi \tag{4.29}$$

$$G = \mu N = NkT \ln(\lambda^3 z) \tag{4.30}$$

The relations are exactly similar to those obtained in the classical model.

GENERIC DISTRIBUTION FUNCTION

The grand canonical s-particle density distribution function, which is the probability of finding s-distinct particles, one in $d\vec{r}_1$ at $\vec{r}_1$, another in $d\vec{r}_2$ at $\vec{r}_2, \ldots$, irrespective of the configuration of the remaining $(N - s)$ particles, is obtained by properly averaging over N

$$n_s(\vec{r}_1, \ldots, \vec{r}_s) = \frac{1}{\Xi} \sum_{N \geq 0} \frac{z^N}{(N-s)!} \int_V \ldots \int e^{-U(\vec{r}_1, \ldots, \vec{r}_N)/kT} \prod_{i=s+1}^{N} d\vec{r}_i \tag{4.31}$$

For the quantal systems, $e^{-U/kT}$ is replaced by Slater sum S_N to give

$$n_s(\vec{r}_1, \ldots, \vec{r}_s) = \frac{1}{\Xi} \sum_{N \geq 0} \frac{z^N}{(N-s)!} \int_V \ldots \int S_N \prod_{i=s+1}^{N} d\vec{r}_i \tag{4.32}$$

The s-particle correlation function $g_s(r_1, \ldots, r_s)$ is defined by

$$n_s(\vec{r}_1, \ldots, \vec{r}_s) = n_1(\vec{r}_1)\, n_1(\vec{r}_2) \ldots n_1(\vec{r}_s)\, g_s(\vec{r}_1, \ldots, \vec{r}_s) \tag{4.33}$$

In a fluid, this reduces to

$$n_s(\vec{r}_1, \ldots, \vec{r}_s) = \rho^s g_s(\vec{r}_1, \ldots, \vec{r}_s) \tag{4.34}$$

Very general physical considerations lead to the following characterization of the distribution functions:

(a) Since ρ_s and n_s are probability densities

$$\rho_s(\vec{r}_1, \ldots, \vec{r}_s) \geq 0$$

$$n_s(\vec{r}_1, \ldots, \vec{r}_s) \geq 0 \tag{4.35}$$

(b) Since the wave function of system of identical particles is either symmetric or antisymmetric, the distribution function exhibits the property of complete symmetry:

$$\rho_s(\vec{r}_1, \ldots, \vec{r}_j, \vec{r}_i, \ldots, \vec{r}_s) = \rho_s(\vec{r}_1, \ldots, \vec{r}_i, \vec{r}_j, \ldots, \vec{r}_s)$$

$$n_s(\vec{r}_1, \ldots, \vec{r}_j, \vec{r}_i, \ldots, \vec{r}_s) = n_s(\vec{r}_1, \ldots, \vec{r}_i, \vec{r}_j, \ldots, \vec{r}_s) \tag{4.36}$$

(c) An arbitrary displacement of all particles does not alter it:

$$\rho_s(\vec{r}_1 + \vec{l}, \ldots, \vec{r}_s + \vec{l}) = \rho_s(\vec{r}_1, \ldots, \vec{r}_s)$$

$$n_s(\vec{r}_1 + \vec{l}, \ldots, \vec{r}_s + \vec{l}) = n_s(\vec{r}_1, \ldots, \vec{r}_s) \tag{4.37}$$

(d) Strong, eventually infinite repulsive interactions between two particles, when they approach closely, lead to

$$\left.\begin{aligned} \rho_s(\vec{r}_1,\ldots,\vec{r}_s) &= 0 \\ n_s(\vec{r}_1,\ldots,\vec{r}_s) &= 0 \end{aligned}\right\} \quad \text{if any } r_{ij} < a \tag{4.38}$$

where a is the hard core of the two particle interaction.

(e) Absence of long-range order of interparticle interaction leads to

$$\left.\begin{aligned} \rho_2(\vec{r}_1,\vec{r}_2) &= \rho^2[1 + O(1/N)] \\ n_s(\vec{r}_1,\vec{r}_2) &= \rho^2 \end{aligned}\right\} \quad \text{if } r_{12} \to \infty \tag{4.39}$$

COMPARISON OF VARIOUS ENSEMBLES

We have considered three types of ensembles. The microcanonical ensemble corresponds to closed isothermic systems, i.e., systems that have energy E, number of particles N (and hence the mass), volume V and temperature T all constant. The canonical ensemble corresponds to isothermic systems that can exchange energy but not the number of particles (hence the mass). Thus, N, V and T are all constant. The grand canonical ensemble corresponds to open isothermic systems, which can exchange both the energy and the number of particles. In this case, the chemical potential μ, V and T are all constant. Other type of ensemble is also possible, which we have not discussed. We have obtained both the canonical distribution and the grand canonical distribution from microcanonical distribution. We can thus say that the microcanonical distribution is fundamental to the basic theory. The thermodynamic quantities of the system can be determined by using any of the ensembles. However, the mathematical calculations in the microcanonical ensemble are most tedious and awkward to handle. Most of the calculations are very conveniently done in the canonical ensemble. Only when the number of particles is changing, one goes to grand canonical ensemble. The factorizability of the partition function in grand canonical ensemble is an advantage.

EXERCISES

1. Derive the Gibbs grand canonical distribution.
2. What is the grand partition function? How are thermodynamic functions related to it?
3. Apply the grand canonical distribution to a perfect gas and obtain Maxwell–Boltzmann distribution.

   Hint: The number of particles in phase volume $dp\,dq$ is $dN = n(p, q)\,dp\,dq$, where $n(p, q)$ is the density in the phase space. Since the gas is dilute, $\langle n\rangle \ll 1$. Replacing N by n and E by ε, the energy of a particle; we have

$$f_0 = e^{\Omega/kT}, \quad f_1 = h^{-3}\, e^{(\Omega+\mu-\varepsilon)/kT}$$

and as $\langle n \rangle <<1$; f_0 is nearly unity, f_1 is small and we put f_2, f_3, etc. equal to zero. We have

$$\langle n \rangle = \sum_n \frac{n\,dw_n}{dp\,dq} = \sum_n n f_n = 0 \times f_0 + 1 \times f_1 = f_0 e^{(\mu-\varepsilon)/kT}$$

$$= e^{(\mu-\varepsilon)/kT},$$ which is the Maxwell–Boltzmann distribution.

4. Find the grand partition function of an ideal gas and hence obtain its equation of state.

Hint: $\Xi = \sum \lambda^{3N} z^N Q_N = \sum \frac{(\lambda^3 z q)^N}{N!} = e^{\lambda^3 zq} = \exp\,(q\,e^{\mu/kT})$

$$\Omega = -kT \ln \Xi = -kT\,\lambda^3 zq = -kTe^{\mu/kT}\left(\frac{2\pi mkT}{h^2}\right)^{3/2} V$$

$$N = -\left(\frac{\partial \Omega}{\partial \mu}\right)_{V,T} = e^{\mu/kT}\left(\frac{2\pi mkT}{h^2}\right)^{3/2} V$$

$$P = -\left(\frac{\partial \Omega}{\partial V}\right)_{\mu,T} = kTe^{\mu/kT}\left(\frac{2\pi mkT}{h^2}\right)^{3/2} = \frac{NkT}{V}$$

5. Prove

(a) $E = \mu N - \dfrac{\partial \ln \Xi}{\partial \beta}$

(b) $N = z - \left(\dfrac{\partial \ln \Xi}{\partial z}\right)_{V,T}$

Hint:

(a) $$E = \langle E \rangle = \sum_{N \ge 0} \frac{1}{N!h^{3N}\Xi} \int E_{nN} e^{\beta(\mu N - E_{nN})}\, dp^N dq^N$$

$$= \sum_{N\ge 0} \frac{1}{N!h^{3N}\,\Xi}\left[-\int (\mu N - E_{nN})\, e^{\beta(\mu N - E_{nN})} dp^N dq^N + \mu \int N e^{\beta(\mu N - E_{nN})} dp^N dq^N\right]$$

$$= -\frac{1}{\Xi}\frac{\partial}{\partial \beta}\frac{1}{h^{3N}\,N!}\int e^{\beta(\mu N - E_{nN})} dp^N dq^N + \mu\langle N \rangle = -\frac{\partial \ln \Xi}{\partial \beta} + \mu N$$

(b) $$N = \langle N \rangle = \sum_{N\ge 0} \frac{1}{N!h^{3N}\Xi}\int N\, e^{\beta(\mu N - E_{nN})}\, dp^N dq^N$$

$$= \frac{1}{\beta\Xi}\sum \frac{1}{N!h^{3N}}\frac{\partial}{\partial \mu}\int e^{\beta(\mu N - E_{nN})} dp^N dq^N$$

$$= \frac{1}{\beta\Xi}\frac{\partial \Xi}{\partial \mu} = \frac{1}{\beta}\frac{\partial \ln \Xi}{\partial \mu} = \frac{1}{\beta}\frac{\partial \ln \Xi}{\partial z}\frac{\partial z}{\partial \mu} = z\frac{\partial \ln \Xi}{\partial z}$$

5 Fluctuations

THE GAUSSIAN DISTRIBUTION

The physical quantities (thermodynamic functions) that describe a macroscopic body in equilibrium approximate their mean values. In equilibrium statistical mechanics, the most probable distribution is based on the fact that the deviations are negligible, thus justifying the use of any ensemble to calculate the thermodynamic functions. But fluctuations become important if they are not thermodynamically negligible order of magnitude, as happens in the neighbourhood of a critical point or two-phase (multi-phase) region fluctuations or the thermodynamic properties that are related to the fluctuations as happens in the case of light scattering from multi-component systems. We begin by finding the probability distribution of these deviations.

The average of deviation $\langle \Delta x \rangle$ of a physical quantity x from its average value $\langle x \rangle$ is zero, i.e.,

$$\langle \Delta x \rangle = \langle x - \langle x \rangle \rangle = \langle x \rangle - \langle x \rangle = 0$$

So we take the mean square deviation $\langle \Delta x^2 \rangle = \langle (x - \langle x \rangle)^2 \rangle$ as a measure of the fluctuations. We have

$$\langle \Delta x^2 \rangle = \langle (x - \langle x \rangle)^2 \rangle = \langle x^2 \rangle + \langle x \rangle^2 - 2\langle x \rangle \langle x \rangle = \langle x^2 \rangle - \langle x \rangle^2$$

Standard deviation is $\delta x = \sqrt{\langle x^2 \rangle - \langle x \rangle^2}$ and standard relative deviation is $\sqrt{(\langle x^2 \rangle - \langle x \rangle^2)/\langle x \rangle^2}$.

Let us assume that the mean value $\langle x \rangle$ has bean subtracted from the measurements x. This amounts to the assumption $\langle x \rangle = 0$ and is simply a matter of shifting of the origin.

We recall from Eq. (1.27) that the probability distribution for exact values of energies of the subsystem is proportional to $e^{\sigma} = e^{S/k}$, with entropy S regarded as a function of energy. Since no specific property of energy is used, similar arguments can be used for the probability $w(x)\,dx$ of a quantity x to have a value between x and $x + dx$, proportional to $e^{S(x)/k}$, i.e.,

$$w(x)\,dx = \text{constant } e^{S(x)/k} \tag{5.1}$$

where $S(x)$ is entropy, formally regarded as a function of x. Einstein first used the above formula in 1907 for the study of fluctuations.

The entropy has a maximum for $x = \langle x \rangle = 0$. Hence $(\partial S/\partial x) = 0$ and $(\partial^2 S/\partial x^2) < 0$ for $x = 0$. For small fluctuations, x is small and we make use of Taylor expansion and retain terms up to second order only to write

$$\frac{S(x) - S(0)}{k} = -\frac{1}{2}\beta x^2 \tag{5.2}$$

We substitute Eq. (5.2) in Eq. (5.1) and treat $e^{-S(0)/k}$ as a constant to get

$$w(x)\,dx = \text{constant } e^{-(1/2)\beta x^2}\,dx$$

The constant is obtained to be $\sqrt{\beta/2\pi}$ from the normalization condition

$$\int_{-\infty}^{\infty} w(x)\,dx = 1$$

We thus have

$$w(x)\,dx = \sqrt{\frac{\beta}{2\pi}}\,e^{-(1/2)\beta x^2}\,dx \tag{5.3}$$

The above is called a *Gaussian distribution*. It has a maximum when $x = 0$ (equivalent to $x = \langle x \rangle$) and decreases rapidly and symmetrically as x increases on either side of the maximum. The mean square fluctuation is

$$\langle x^2 \rangle = \sqrt{\frac{\beta}{2\pi}} \int_{-\infty}^{\infty} x^2 e^{-(1/2)\beta x^2}\,dx = \frac{1}{\beta} \tag{5.4}$$

This gives for the Gaussian distribution

$$w(x)\,dx = \frac{1}{\sqrt{2\pi\langle x^2 \rangle}} \exp\left(\frac{-x^2}{2\langle x^2 \rangle}\right) dx \tag{5.5}$$

As $\langle x^2 \rangle$ becomes smaller, the maximum becomes sharper; $w(x)$ becomes virtually a Dirac-delta function.

If we do not take $\langle x \rangle = 0$, then the mean square function $\langle \Delta x^2 \rangle = \langle x^2 \rangle - \langle x \rangle^2$, and we have to replace $\langle x^2 \rangle$ by $\langle x^2 \rangle - \langle x \rangle^2$ in Eq. (5.5) to write the Gaussian distribution as

$$w(x)\,dx = \frac{1}{\sqrt{2\pi(\langle x^2 \rangle - \langle x \rangle^2)}} \exp\left[\frac{(x - \langle x \rangle)^2}{2(\langle x^2 \rangle - \langle x \rangle^2)}\right] dx \tag{5.6}$$

Ordinarily, for large thermodynamic systems, small fluctuations can be legitimately represented by a Gaussian distribution (Fig. 5.1).

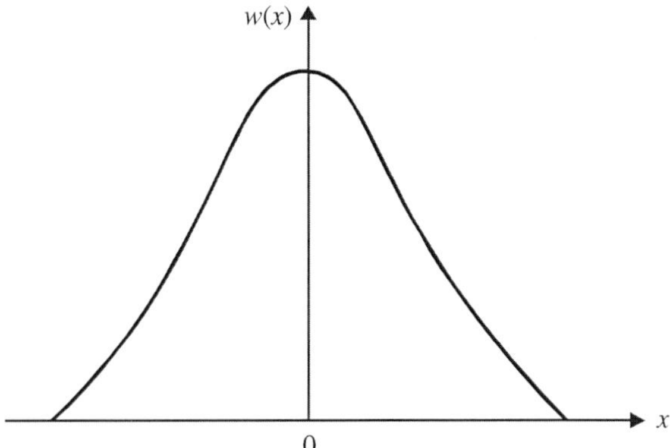

Figure 5.1 Gaussian distribution.

FLUCTUATIONS IN VARIOUS ENSEMBLES

We shall now examine the fluctuations of the energy in canonical ensemble and that of both, the energy and the number of particles in grand canonical ensemble. If the fluctuations of energy in the canonical ensemble is negligible, it corresponds to microcanonical ensemble and if the fluctuations in number of particles in a grand canonical ensemble is negligible, it corresponds to a canonical ensemble. Thus, the negligibility of fluctuations establishes the equivalence of ensembles.

Canonical Ensemble

We evaluate below the standard relative deviation of energy that will be a measure of the fluctuation. We have

$$\langle E \rangle = \frac{1}{Q_N} \sum_n E_n e^{-E_n/kT} \tag{5.7}$$

Differentiating with respect to temperature, with the volume and the number of particles kept constant, we obtain

$$\left(\frac{\partial \langle E \rangle}{\partial T}\right)_{N,V} Q_N + \langle E \rangle \left(\frac{\partial Q_N}{\partial T}\right)_{N,V} = \frac{\partial}{\partial T}\left(\sum_n E_n e^{-E_n/kT}\right)_{N,V}$$

or

$$\left(\frac{\partial \langle E \rangle}{\partial T}\right)_{N,V} Q_N + \frac{\langle E \rangle}{kT^2} \sum_n E_n e^{-E_n/kT} = \frac{1}{kT^2} \sum_n E_n^2 e^{-E_n/kT}$$

The above simplifies to

$$\frac{\langle E^2 \rangle - \langle E \rangle^2}{kT^2} = \left(\frac{\partial \langle E \rangle}{\partial T}\right)_{N,V} \tag{5.8}$$

As thermodynamically $E \leftrightarrow \langle E \rangle$ and $(\partial E/\partial T)_V = C_V$, we have

$$\frac{\langle E^2 \rangle - \langle E \rangle^2}{\langle E \rangle^2} = \frac{kT^2 C_V}{E^2} \tag{5.9}$$

Since $E = O(N\,kT)$ and $C_V = O(N\,k)$, hence

$$\frac{\langle E^2 \rangle - \langle E \rangle^2}{\langle E \rangle^2} = O\left(\frac{1}{N}\right) \tag{5.10}$$

As $N \sim 10^{23}$ for a mole, the standard relative deviation of energy is of the order of $1/\sqrt{N}$, i.e. $O(10^{-11})$. We therefore conclude that the fluctuations in energy are extremely small, i.e. the distribution in energy is extremely sharp peaked and the ensemble is practically a microcanonical ensemble.

Using Eq. (1.39), let us now consider the ensemble average of pressure.

$$\langle P \rangle Q_N = \sum_n \left(-\frac{\partial E_n}{\partial V}\right)_S e^{-E_n/kT} \tag{5.11}$$

Differentiating with respect to V, with N and T constant, we get

$$\left(\frac{\partial \langle P \rangle}{\partial V}\right)_{N,T} Q_N + \frac{\langle P \rangle}{kT} \sum_n \left(-\frac{\partial E_n}{\partial V}\right)_{N,T} e^{-E_n/kT} = \sum_n \left(-\frac{\partial^2 E_n}{\partial V^2}\right) e^{-E_n/kT} + \frac{1}{kT} \sum_n \left(\frac{\partial E_n}{\partial V}\right)^2 e^{-E_n/kT}$$

Hence

$$\left(\frac{\partial \langle P \rangle}{\partial V}\right)_{N,T} + \frac{\langle P \rangle^2}{kT} = \frac{\langle P^2 \rangle}{kT} - \left\langle \frac{\partial^2 E}{\partial V^2} \right\rangle$$

This yields

$$\langle P^2 \rangle - \langle P \rangle^2 = kT \left[\left(\frac{\partial \langle P \rangle}{\partial V}\right)_{N,T} + \left\langle \frac{\partial^2 E}{\partial V^2} \right\rangle\right] \tag{5.12}$$

With the thermodynamic association $P \leftrightarrow \langle P \rangle$, we have

$$\frac{\langle P^2 \rangle - \langle P \rangle^2}{\langle P \rangle^2} = \frac{kT}{P^2} \left[\left(\frac{\partial P}{\partial V}\right)_{N,T} - \left\langle \frac{\partial^2 E}{\partial V^2} \right\rangle\right] \tag{5.13}$$

The first term on right-hand side is $O(-1/N)$ since $PV = NkT$ and $(\partial P/\partial V)_{N,T} = -P^2/NkT$. The second term can be evaluated with the help of the expression for energy $\varepsilon = (n^2\pi^2\hbar^2)/2V^{2/3}$ of a particle in cubic box of volume V and $E = N\varepsilon$. This gives $\langle \partial^2 E/\partial V^2 \rangle = (5/3)(P^2/NkT)$. Thus the standard relative deviation is of the order of $N^{-1/2}$ and hence negligible.

Grand Canonical Ensemble

We evaluate the standard relative deviation of number of particles. We start with

$$\langle N \rangle = \frac{1}{\Xi} \sum_{n,N} N e^{(\mu N - E_{nN})/kT} \tag{5.14}$$

Differentiating with respect to μ, we have

$$\left(\frac{\partial\langle N\rangle}{\partial\mu}\right)_{V,T}\Xi+\frac{\langle N\rangle}{kT}\sum_{n,N}Ne^{(\mu N-E_{nN})/kT}=\frac{1}{kT}\sum_{n,N}N^2e^{(\mu N-E_{nN})/kT}$$

(It should be noted that N is not a function of μ but $\langle N\rangle$ is.)

This gives

$$\left(\frac{\partial\langle N\rangle}{\partial\mu}\right)_{V,T}=\frac{\langle N^2\rangle-\langle N\rangle^2}{kT} \tag{5.15}$$

From Gibbs–Duhem Eq. (1.69), we have for constant T with $V/N = v$:

$$d\mu = v\,dP$$

or

$$\left(\frac{\partial\mu}{\partial v}\right)_T=v\left(\frac{\partial P}{\partial v}\right)_T \tag{5.16}$$

Also,

$$\frac{\partial\mu}{\partial N}=\frac{\partial\mu}{\partial v}\frac{\partial v}{\partial N}=v\left(\frac{\partial P}{\partial v}\right)_T\frac{\partial v}{\partial N}=v\frac{\partial P}{\partial v}\left(-\frac{V}{N^2}\right)=-\frac{V^2}{N^2}\frac{dP}{dV}$$

Hence

$$\left(\frac{\partial N}{\partial\mu}\right)_{V,T}=-\frac{N^2}{V^2}\left(\frac{\partial V}{\partial P}\right)_{N,T}=\frac{N^2}{V}K_T \tag{5.17}$$

where $K_T=-\frac{1}{V}\left(\frac{\partial V}{\partial P}\right)_{N,T}$ is isothermal compressibility.

Substitution from Eq. (5.17) in Eq. (5.15) gives

$$\frac{\langle N^2\rangle-\langle N\rangle^2}{\langle N\rangle^2}=\frac{kT}{V}K_T=\frac{\langle\rho^2\rangle-\langle\rho\rangle^2}{\langle\rho\rangle^2} \tag{5.18}$$

where ρ (= N/V) is the number density.

For an ideal gas,

$$\frac{kT}{V}K_T=-\frac{kT}{V^2}\left(\frac{\partial V}{\partial P}\right)_{N,T}=O\left(\frac{1}{N}\right)$$

Thus the standard relative deviation is of the order of $N^{-1/2}$. The fluctuation is significant near the liquid gas critical point as K_T tends to infinity. These fluctuations then give rise to large scattering of light, known as *opalescence of light.*

To find the fluctuation in pressure, we begin with

$$\langle P\rangle\,\Xi=-\sum_{n,N}\left(-\frac{\partial E_{nN}}{\partial V}\right)e^{(\mu N-E_{nN})/kT} \tag{5.19}$$

Differentiating with respect to V with μ and T constant, we obtain as in the canonical case

$$\langle P^2 \rangle - \langle P \rangle^2 = kT \left[\left(\frac{\partial \langle P \rangle}{\partial V} \right)_{\mu,T} - \left\langle \frac{\partial P}{\partial V} \right\rangle \right] \tag{5.20}$$

Since P is a function of μ and T only, it reduces to

$$\frac{\langle P^2 \rangle - \langle P \rangle^2}{\langle P \rangle^2} = -\frac{kT}{P^2} \left\langle \frac{\partial P}{\partial V} \right\rangle \tag{5.21}$$

For the fluctuation in energy, we recall that $E \leftrightarrow \langle E \rangle \equiv \langle E_{nN} \rangle$ and start with

$$\langle E \rangle = \frac{\sum\limits_{n,N} E_{nN}\, e^{\beta(\mu N - E_{nN})}}{\sum\limits_{n,N} e^{\beta(\mu N - E_{nN})}} \tag{5.22}$$

Differentiating with respect to β and keeping μ and V constant, we have

$$\begin{aligned} \frac{\partial \langle E \rangle}{\partial \beta} &= \frac{\sum E_{nN} (\mu N - E_{nN})\, e^{\beta(\mu N - E_{nN})}}{\sum e^{\beta(\mu N - E_{nN})}} - \frac{\sum E_{nN}\, e^{\beta(\mu N - E_{nN})}}{\sum e^{\beta(\mu N - E_{nN})}} \frac{\sum (\mu N - E_{nN}) e^{\beta(\mu N - E_{nN})}}{\sum e^{\beta(\mu N - E_{nN})}} \\ &= \mu \langle EN \rangle - \langle E^2 \rangle - \mu \langle E \rangle \langle N \rangle + \langle E \rangle^2 \\ &= \langle E \rangle^2 - \langle E^2 \rangle + \mu \left(\langle EN \rangle - \langle E \rangle \langle N \rangle \right) \end{aligned} \tag{5.23}$$

Differentiating Eq. (5.22) with respect to μ and keeping V and T constant, we have

$$\begin{aligned} \frac{\partial \langle E \rangle}{\partial \mu} &= \frac{\sum E_{nN} \beta N\, e^{\beta(\mu N - E_{nN})}}{\sum e^{\beta(\mu N - E_{nN})}} - \frac{\sum E_{nN}\, e^{\beta(\mu N - E_{nN})}}{\sum e^{\beta(\mu N - E_{nN})}} \frac{\sum \beta N e^{\beta(\mu N - E_{nN})}}{\sum e^{\beta(\mu N - E_{nN})}} \\ &= \beta \left(\langle EN \rangle - \langle E \rangle \langle N \rangle \right) \end{aligned} \tag{5.24}$$

From Eqs. (5.23) and (5.24), we obtain

$$\langle E^2 \rangle - \langle E \rangle^2 = kT^2 \frac{\partial E}{\partial T} + \mu kT \frac{\partial E}{\partial \mu} = kT^2 \left(\frac{\partial E}{\partial T} \right) + \mu kT \frac{\partial E}{\partial N} \frac{\partial N}{\partial \mu} \tag{5.25}$$

If we examine Eq. (5.25) together with Eqs. (5.8) and (5.15), we observe that the mean square deviation of energy in grand canonical ensemble is made up of the sum of two terms, the first one being the mean square deviation of energy in canonical ensemble and the second term arising due to the mean square deviation of the number of particles. Further, the fluctuations in E and N are not independent of each other and are correlated.

THERMODYNAMIC EQUIVALENCE OF ENSEMBLES

We have seen that the fluctuation in a thermodynamic property is $O(1/N)$. Hence in the formulation of thermodynamics, fluctuations can be neglected. Large fluctuations do occur in two-phase and critical regions, but are consistent with thermodynamics and pose no problems, and so need not be mentioned here.

In practice, the thermodynamic deductions from statistical mechanics depend only on the mean values. This eliminates the uniqueness of the *proper ensemble* in a given application and we thus have thermodynamic equivalence which we illustrate here.

Canonical and Microcanonical Ensembles

For a canonical ensemble,

$$Q_N = \sum_E \Delta\Gamma(E, N, V)\, e^{-E/kT} \tag{5.26}$$

For large values of N, the successive values of E are so close together that the summation can be replaced by an integration

$$Q_N = \int d\Gamma(E,N,V)\, e^{-E/kT} \tag{5.27}$$

(in the classical case, the above is the expression).

If the fluctuations over energy in the canonical ensemble are negligible, the main contribution to the integral comes when the integrand is maximum, which will correspond to $E = E^*$ (say), determined by

$$\left(\frac{\partial(\Delta\Gamma e^{-E/kT})}{\partial E}\right)_{N,V,T} = 0$$

This gives

$$\left(\frac{\partial \ln \Delta\Gamma}{\partial E}\right)_{N,V} = \frac{1}{kT} \tag{5.28}$$

We then have

$$Q_N = \Delta\Gamma(E^*, N, V)\, e^{-E^*/kT}$$

This gives

$$\frac{P}{kT} = \left(\frac{\partial \ln Q_N}{\partial V}\right)_{N,T} = \left(\frac{\partial \ln \Delta\Gamma}{\partial V}\right)_{N,E=E^*} \tag{5.29}$$

and

$$S = k \ln Q_N + kT\left(\frac{\partial \ln Q_N}{\partial T}\right)_{V,N} = k \ln \Delta\Gamma(E^*, N, V) \tag{5.30}$$

which are same as the microcanonical ensemble equations. This establishes the equivalence.

Grand Canonical and Canonical Ensemble

We begin with

$$\Xi(\mu,V,T) = \sum_N e^{\mu N/kT} Q_N(N, V, T) \tag{5.31}$$

and the largest term occurs for $N = N^*$ determined by

$$\left(\frac{\partial(e^{\mu N/kT}Q_N}{\partial N}\right)_{\mu,V,T} = 0$$

This gives

$$\mu = -kT\left(\frac{\partial \ln Q_N}{\partial N}\right)_{V,T} = \left(\frac{\partial A}{\partial N}\right)_{V,T} \tag{5.32}$$

The grand canonical partition function is then

$$\Xi = e^{\mu N^*/kT} Q_{N^*}(N^*, V, T)$$

This gives

$$P = kT\left(\frac{\partial \ln \Xi}{\partial V}\right)_{\mu,T} = kT\left(\frac{\partial \ln Q_{N^*}}{\partial V}\right)_T \tag{5.33}$$

$$S = k\ln \Xi + kT\left(\frac{\partial \ln \Xi}{\partial T}\right)_{\mu,V} = k\ln Q_{N^*} + kT\left(\frac{\partial \ln Q_{N^*}}{\partial T}\right)_V \tag{5.34}$$

and for

$$\langle N\rangle = kT\left(\frac{\partial \ln \Xi}{\partial \mu}\right)_{V,T} = N^* \tag{5.35}$$

which are same as the canonical ensemble equations, establishing the equivalence.

So far we have dealt with the fluctuations from the mean values only. The fluctuations taking place with time, give rise in a natural way to transport phenomenon and thus the non-equilibrium statistical mechanics, discussed later.

EXERCISES

1. Calculate the standard relative deviation of energy of a canonical ensemble and examine the condition under which it is equivalent to a microcanonical ensemble.
2. Find the fluctuation of the number of particles in a grand canonical ensemble and establish its equivalence to a canonical ensemble.
3. Show that the mean square deviation of energy in grand canonical ensemble is the sum of the mean square deviation of energy in a canonical ensemble and a term due to the fluctuation of the number of particles.
4. Establish the thermodynamic equivalence of the microcanonical, the canonical and the grand canonical ensembles.

6 Ideal Fermi Gas

THE FERMI–DIRAC DISTRIBUTION

At low temperatures, where the mean occupation number is not small, the Boltzmann statistics cannot be applied. The quantum statistics to be used depends upon the type of wave function, symmetric or antisymmetric, that describes the gas. We consider here, a system of particles described by antisymmetric wave function. The Pauli exclusion principle applies in this case and there can be no more than one particle in any one quantum state. It has already been mentioned that such a system of particles obey the Fermi–Dirac distribution.

We obtain the Fermi–Dirac distribution by applying the Gibbs distribution to the set of all the particles in the gas, which are in the same ith quantum state. From the general formula for the grand potential Eq. (4.24) with n_i particles in the ith quantum state of energy ε_i (the energy of n_i particles will be $n_i\ \varepsilon_i$), we have

$$\Omega_i = -kT \ln \sum_{n_i} (e^{(\mu-\varepsilon_i)/kT})^{n_i} \tag{6.1}$$

For *fermions*, the mean occupation number is either 0 or 1. We then have

$$\Omega_i = -kT \ln [1+e^{(\mu-\varepsilon_i)/kT}] \tag{6.2}$$

Differentiating Eq. (6.2) with respect to μ with V and T constant and using Eq. (4.28), we get

$$\langle n_i \rangle = \frac{e^{(\mu-\varepsilon_i)/kT}}{1+e^{(\mu-\varepsilon_i)/kT}}$$

which reduces to

$$\langle n_i \rangle = \frac{1}{e^{(\varepsilon_i-\mu)/kT}+1} \tag{6.3}$$

Dropping the subscript, we write

$$f = \frac{1}{e^{(\varepsilon-\mu)/kT}+1} \tag{6.4}$$

This is the distribution function for an ideal Fermi gas. We see that $<n> \leq 1$, which is the expected value. The Fermi distribution is normalized by the condition

$$\sum_i \frac{1}{e^{(\varepsilon_i-\mu)/kT}+1} = N \tag{6.5}$$

where N is the total number of particles in the gas. This equation also determines μ as a function of N and T.

THE DEGENERATE ELECTRON GAS

Let us consider an electron gas. We shall deal with two cases:

(a) at absolute zero (T = 0), the electron gas is completely degenerate and

(b) at low temperatures ($T < T_F$), the gas is degenerate.

At Absolute Zero

The gas is completely degenerate, all energy levels up to some largest value is completely filled up with two electrons (spin up and spin down).

The number of states in μ space of an element of volume $d\vec{p}\cdot d\vec{r}$ is

$$g\,\frac{d\vec{p}\cdot d\vec{r}}{h^3}$$

where $g = (2s + 1)$, is called spin degeneracy and s is the spin quantum number of the particle. For electrons, $g = 2$.

The number of quantum states with momentum between $\vec{p}$ and $\vec{p} + d\vec{p}$ is obtained, by integrating the above over $d\vec{r}$ that yields V, the volume of the gas, and putting $d\vec{p} \equiv 4\pi\, p^2\, dp$ (spherically symmetric case) as:

$$g(p)dp = \frac{4\pi g V p^2 dp}{h^3} = \frac{2\pi g V}{h^3}\,(2m)^{3/2}\,\varepsilon^{1/2}\,d\varepsilon \equiv g(\varepsilon)d\varepsilon \tag{6.6}$$

At absolute zero, the electrons are distributed over the different quantum states in such a way that the total energy of the gas is minimum. Because of Pauli principle, there cannot be more than one electron in any one quantum state. Also recall that a quantum state is characterized by the quantum numbers n, l, m and s, whereas the energy level is characterized by quantum numbers n, l and m only. Hence, there can be two electrons (one spin up and another spin down) in each energy level. In these circumstances, all energy states starting from zero up to some largest value, determined by the total number of electrons in the gas, will be filled up. Hence we have

$$N = \frac{4\pi g V}{h^3}\int_0^{p_F} p^2\,dp \tag{6.7}$$

where p_F is the value of the momentum corresponding to the largest energy value, and is called Fermi momentum. This gives

$$p_F = h\left(\frac{3}{4\pi g}\cdot\frac{N}{V}\right)^{1/3} \tag{6.8}$$

and the corresponding limiting value of energy ε_F is given by

$$\varepsilon_F = \frac{h^2}{2m}\left(\frac{3}{4\pi g}\frac{N}{V}\right)^{2/3} \tag{6.9}$$

This energy is called the Fermi energy. The corresponding level is Fermi level and p_F is the radius of the Fermi sphere in the momentum space.

To examine the thermodynamic meaning of the Fermi energy, let us consider the Fermi–Dirac distribution function

$$f = \frac{1}{e^{(\varepsilon-\mu)/kT}+1} \tag{6.10}$$

Here we have dropped the subscript.

$$\begin{aligned} &\text{For } \varepsilon > \mu,\ f \to 0 \text{ as } T \to 0 \text{ K} \\ &\text{For } \varepsilon < \mu,\ f \to 1 \text{ as } T \to 0 \text{ K} \end{aligned} \tag{6.11}$$

This is represented by the continuous line in the Fig. 6.1.

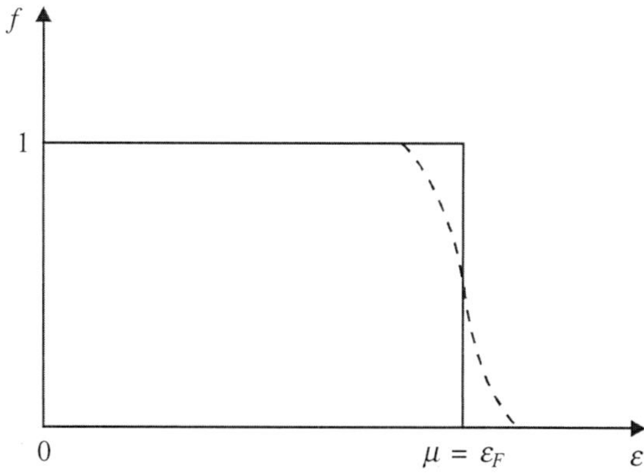

Figure 6.1 The Fermi–Dirac distribution function.

From this, one concludes that at absolute zero, the chemical potential of the gas coincides with the Fermi energy:

$$\mu = \varepsilon_F \quad \text{at} \quad T = 0 \text{ K} \tag{6.12}$$

We thus see that the Fermi energy is the cut-off energy. At absolute zero, all energy levels up to ε_F are completely filled by electrons and all energy levels above it are unoccupied.

The temperature determined by the relation $\varepsilon_F = kT_F$ is called the Fermi temperature. From Eq. (6.9) we find for electrons

$$\varepsilon_F = 0.625 \times 10^{-17} \rho^{2/3} \text{ J} = 39 \rho^{2/3} \text{ eV}$$

where the density of the gas $\rho = Nm/V$ kg/m^3.

The total energy of the gas is obtained by multiplying the number of states by $p^2/2m$ and integrating over p from zero to p_F, i.e.

$$E = \frac{4\pi gV}{2mh^3}\int_0^{p_F} p^4\,dp = \frac{4\pi gV}{2mh^3}\cdot\frac{p_F^5}{5} = \frac{3}{5}N\varepsilon_F \tag{6.13}$$

The equation of state is obtained by using $PV = 2/3E$ as

$$P = \frac{1}{5}\frac{h^2}{m}\left(\frac{3}{4\pi g}\right)^{2/3}\left(\frac{N}{V}\right)^{5/3} \tag{6.14}$$

We thus see that a Fermi gas even at absolute zero possesses energy and exerts pressure.

It is easily seen that μ is positive (though small) at $T = 0$ K, decreases as temperature increases and is negative at high temperatures. The coulomb attraction between the ions (in metals) counterbalances the pressure.

For a complete degenerate extreme relativistic electron gas (at high densities, i.e., for a very compressed gas, the energy of the electron gas increases and when the energy becomes comparable with mc^2, the relativistic case arises), we have

$$\varepsilon = cp$$

and Eqs. (6.6) and (6.8) still holding, give for the Fermi energy

$$\varepsilon_F = cp_F = ch\left(\frac{3}{4\pi g}\frac{N}{V}\right)^{1/3} \tag{6.15}$$

The total energy of the gas is

$$E = \frac{4\pi gcV}{h^3}\int_0^{p_F} p^3\,dp = \frac{\pi g}{h^3}cp_F^4 V$$

$$= \frac{3}{4}chN\left(\frac{3}{4\pi g}\frac{N}{V}\right)^{1/3} \tag{6.16}$$

The pressure of the gas is obtained by using

$$P = -\left(\frac{\partial E}{\partial V}\right)_S$$

This gives

$$P = \frac{1}{4}ch\left(\frac{3}{4\pi g}\right)^{1/3}\left(\frac{N}{V}\right)^{4/3} = \frac{1}{3}\frac{E}{V} \tag{6.17}$$

with $g = 2$ for electron gas. It should be noted that $PV = 1/3\ E$ for an extreme relativistic gas in contrast to the non-relativistic case $PV = 2/3E$.

At Low Temperatures ($T < T_F$)

At low temperatures, the changes in distribution function take place chiefly when close to ε_F. This is represented by the dotted line in Fig. 6.1. We shall make use of this fact to make a Taylor expansion.

First let us consider an integral of the form

$$I = \int_0^{\infty} \frac{\phi(\varepsilon)}{e^{(\varepsilon-\mu)/kT} + 1} d\varepsilon \tag{6.18}$$

where $\phi(\varepsilon)$ is some function of ε such that the integral converges.

If we put

$$\frac{\varepsilon - \mu}{kT} = x \tag{6.19}$$

the integral I in Eq. (6.18) becomes

$$I = kT \int_{-\mu/kT}^{\infty} \frac{\phi(\mu + kTx)}{e^x + 1} dx \tag{6.20}$$

We break the integral on the right-hand side in two parts:

$$I = kT \int_{-\mu/kT}^{0} \frac{\phi(\mu+kTx)}{e^x + 1} dx + kT \int_0^{\infty} \frac{\phi(\mu+kTx)}{e^x + 1} dx \tag{6.21}$$

In the first integral on right-hand side, if we invert the limits of integration and replace x by $-x$, the integral is unchanged. We then use the identity

$$\frac{1}{e^{-x} + 1} = 1 - \frac{1}{e^x + 1}$$

to obtain

$$I = \int_0^{\mu} \phi(\varepsilon) d\varepsilon - kT \int_0^{\mu/kT} \frac{\phi(\mu - kTx)}{e^x + 1} dx + kT \int_0^{\infty} \frac{\phi(\mu+kTx)}{e^x + 1} dx \tag{6.22}$$

Since $\mu/kT \gg 1$ and the integral converges rapidly, we can replace the upper limit in the second integral by infinity, without much error, to get

$$I = \int_0^{\mu} \phi(\varepsilon)\, d\varepsilon + kT \int_0^{\infty} \frac{\phi(\mu+kTx) - \phi(\mu-kTx)}{e^x + 1} dx$$

We now use the Taylor expansion

$$\phi(\mu \pm kTx) = \phi(\mu) + (\pm kTx)\, \phi'(\mu) + \frac{(\pm kTx)^2}{2} \phi''(\mu) + \cdots$$

to get

$$I = \int_0^{\mu} \phi(\varepsilon)\, d\varepsilon + 2(kT)^2\, \phi'(\mu) \int_0^{\infty} \frac{x\, dx}{e^x + 1} + \frac{1}{3}(kT)^4\, \phi'''(\mu) \int_0^{\infty} \frac{x^3 dx}{e^x + 1} + \cdots \tag{6.23}$$

The integrals of the type involved in Eq. (6.23) are evaluated in the following way:

$$\begin{aligned}
\int_0^\infty \frac{x^{k-1}\,dx}{e^x+1} &= \int_0^\infty dx\, x^{k-1}\, e^{-x}\,(1+e^{-x})^{-1} \\
&= \int_0^\infty dx\, x^{k-1} \sum_{m=1}^{\infty} (-1)^{m+1} e^{-mx} \\
&= \sum_{m=1}^{\infty} \frac{(-1)^{m+1}}{m^k} \int_0^\infty dz\, z^{k-1}\, e^{-z} \quad \text{where } z = mx \\
&= \sum_{m=1}^{\infty} \frac{(-1)^{m+1}}{m^k}\,\Gamma(k) = \Gamma(k)\left[\frac{1}{1^k} - \frac{1}{2^k} + \frac{1}{3^k} - \frac{1}{4^k} + \cdots\right] \\
&= \Gamma(k)\left[\frac{1}{1^k} + \frac{1}{2^k} + \frac{1}{3^k} + \cdots - \frac{2}{2^k}\left(\frac{1}{1^k} + \frac{1}{2^k} + \frac{1}{3^k} + \cdots\right)\right] \\
&= \Gamma(k)(1-2^{1-k})\,\zeta(k) \qquad (6.24)
\end{aligned}$$

where $\Gamma(k)$ and $\zeta(k)$ are gamma and Riemann zeta functions defined by

$$\begin{aligned}
\Gamma(k) &= \int_0^\infty dt\, e^{-t}\, t^{k-1},\ \Gamma(k+1) = k! \ \text{ for } k \text{ an integer} \\
\zeta(k) &= \sum_{m=1}^{\infty} \frac{1}{m^k} \qquad (6.25)
\end{aligned}$$

Tables for values of gamma and zeta functions are available; we quote some of the values

$$\Gamma\left(\frac{1}{2}\right) = \sqrt{\pi},\ \Gamma\left(\frac{3}{2}\right) = \frac{1}{2}\sqrt{\pi},\ \Gamma\left(\frac{5}{2}\right) = \frac{3}{4}\sqrt{\pi}$$

$$\zeta(2) = \left(\frac{\pi^2}{6}\right),\ \zeta(3) = 1.202,\ \zeta(4) = \left(\frac{\pi^4}{90}\right),\ \zeta(5) = 1.037$$

$$\zeta\left(\frac{3}{2}\right) = 2.612,\ \zeta\left(\frac{5}{2}\right) = 1.341$$

Substitution yields the value of the integral

$$\begin{aligned}
I &= \int_0^\infty \frac{\phi(\varepsilon)}{e^{(\varepsilon-\mu)/kT}+1}\,d\varepsilon \\
&= \int_0^\mu \phi(\varepsilon)\,d\varepsilon + \frac{\pi^2}{6}(kT)^2\,\phi'(\mu) + \frac{7\pi^4}{360}(kT)^4\,\phi'''(\mu) + \cdots \qquad (6.26)
\end{aligned}$$

We now evaluate the Fermi energy at low temperatures. To do so, we start by calculating the number of electrons, which expressed in terms of $\varepsilon = p^2/2m$ is given from Eq. (6.8) by

$$N = \frac{4\pi gV}{h^3}\sqrt{2}\, m^{3/2} \int_0^\infty \frac{\varepsilon^{1/2}\, d\varepsilon}{e^{(\varepsilon-\mu)/kT}+1} \tag{6.27}$$

Setting $\phi(\varepsilon) = \varepsilon^{1/2}$, we get with the aid of Eq. (6.26), writing $\mu = \varepsilon_F = \varepsilon_{F_0} + (\varepsilon_F - \varepsilon_{F_0})$

$$N = \frac{4\pi gV}{h^3}\sqrt{2} m^{3/2}\left[\int_0^{\varepsilon_{F_0}} \phi(\varepsilon)\, d\varepsilon + \int_{\varepsilon_{F_0}}^{\varepsilon_F} \phi(\varepsilon)\, d\varepsilon + \frac{\pi^2}{6}(kT)^2 \phi'(\varepsilon_F)\right] \tag{6.28}$$

where ε_{F_0} is the Fermi energy at $T = 0$ K and neglecting terms containing higher powers of T.

The first term on the right-hand side of Eq. (6.28) is the number of electrons at 0 K and is same as N, as it does not change with the temperature. This gives

$$\int_{\varepsilon_{F_0}}^{\varepsilon_F} \phi(\varepsilon)\, d\varepsilon = -\frac{\pi^2}{6}(kT)^2 \phi'(\varepsilon_F)$$

Substituting $\phi(\varepsilon) = \varepsilon^{1/2}$, since ε_{F_0} and ε_F do not differ by much, we have

$$(\varepsilon_F - \varepsilon_{F_0})\,\varepsilon_F^{1/2} = -\frac{\pi^2}{12}(kT)^2 \varepsilon_F^{-1/2} \tag{6.29}$$

This gives

$$\varepsilon_F = \varepsilon_{F_0}\left[1 - \frac{\pi^2}{12}\left(\frac{kT}{\varepsilon_{F_0}}\right)^2\right] \tag{6.30}$$

in which we have used $\varepsilon_F \varepsilon_{F_0} \simeq \varepsilon_{F_0}^2$ in the second smaller term.

We thus see that the Fermi energy decreases slightly as the temperature increases. The energy of the gas is given by

$$E = \frac{4\pi gV}{h^3}\sqrt{2}\, m^{3/2} \int_0^\infty \frac{\varepsilon^{3/2}\, d\varepsilon}{e^{(\varepsilon-\mu)/kT}+1} \tag{6.31}$$

Setting $\phi(\varepsilon) = \varepsilon^{3/2}$, we have with the aid of Eq. (6.26),

$$E = \frac{4\pi gV}{h^3}\sqrt{2}\, m^{3/2}\left[\int_0^{\varepsilon_{F_0}} \phi(\varepsilon)\, d\varepsilon + \int_{\varepsilon_{F_0}}^{\varepsilon_F} \phi(\varepsilon)\, d\varepsilon + \frac{\pi^2}{6}(kT)^2 \phi'(\varepsilon_F)\right] \tag{6.32}$$

where we have neglected terms containing higher powers of T.

The first term on the right-hand side is the energy E_0 at $T = 0$ K. Thus,

$$E = E_0 + \frac{4\pi gV}{h^3}\sqrt{2}\, m^{3/2}\left[(\varepsilon_F - \varepsilon_{F_0})\,\varepsilon_F^{3/2} + \frac{3}{2}\frac{\pi^2}{6}(kT)^2 \varepsilon_F^{1/2}\right]$$

The above with the aid of Eq. (6.29) simplifies to

$$E = E_0 + \frac{4\pi gV}{h^3}\sqrt{2}\, m^{3/2}\,\frac{\pi^2}{6}(kT)^2 \varepsilon_F^{1/2} + \cdots \tag{6.33}$$

This can be written as

$$E = \frac{3}{5} N\varepsilon_F \left[1 + \frac{5\pi^2}{12} \left(\frac{kT}{\varepsilon_F} \right)^2 + \cdots \right] \tag{6.34}$$

and the equation of state is

$$P - \frac{2}{3}\frac{E}{V} = \frac{2}{5}\frac{N}{V} \varepsilon_F \left[1 + \frac{5\pi^2}{12} \left(\frac{kT}{\varepsilon_F} \right)^2 + \cdots \right] \tag{6.35}$$

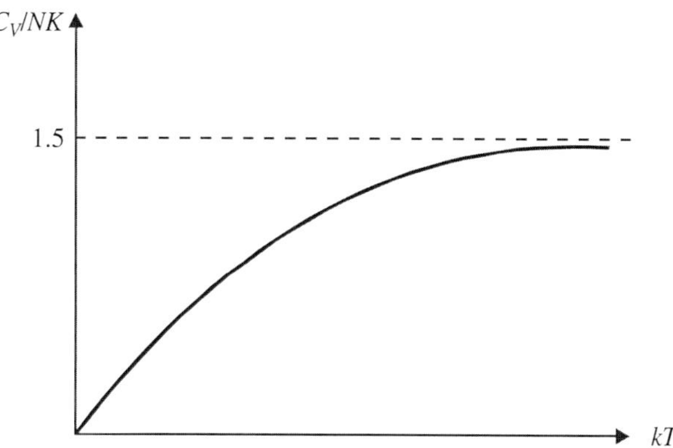

Figure 6.2 Specific heat of an ideal Fermi gas.

The specific heat of the electron gas is

$$C_V = \left(\frac{\partial E}{\partial T} \right)_V = \frac{4\pi g V}{3h^3} (2m\varepsilon_F)^{3/2} \frac{\pi^2}{2} \frac{k^2 T}{\varepsilon_F} = \frac{\pi^2}{2} \frac{Nk^2 T}{\varepsilon_F} = \frac{\pi^2}{2} Nk \frac{T}{T_F} \tag{6.36}$$

Thus, the specific heat is proportional to the temperature at low temperatures (Sommerfield, 1928) and is much smaller than the classical value $(3/2)Nk$ (which is temperature independent) at room temperature (Fig. 6.2). The total specific heat of a metal is the sum of the contribution due to the electronic part (obtained above) and the contribution due to the lattice part at low temperatures. We have

$$C_{V,\text{lat}} = \frac{12}{5} \pi^4 Nk \left(\frac{T}{\Theta_D} \right)^3 \tag{6.37}$$

where Θ_D is the Debye temperature. The contribution due to the lattice part is proportional to T^3 (see Chapter 7). Hence, for the specific heat of a metal at low temperatures, we can write

$$C_V = \gamma T + \alpha T^3$$

where γ and α are constants.

or

$$\frac{C_V}{T} = \gamma + \alpha T^2 \tag{6.38}$$

Kok and Keesom obtained the graph experimentally for copper (Fig. 6.3).

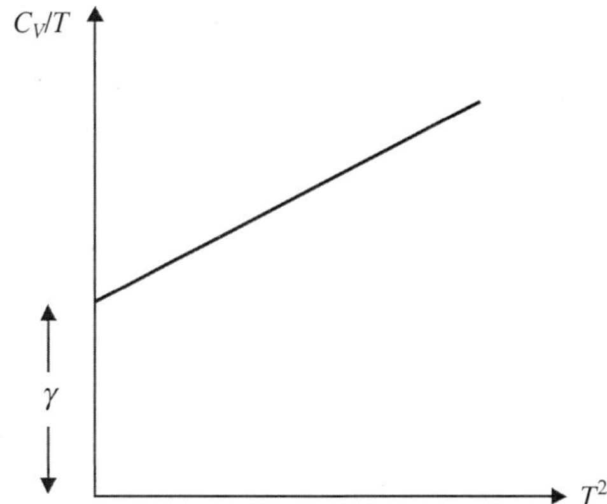

Figure 6.3 Variation of specific heat of copper.

The observed value of γ was found to be higher than the calculated value from Eq. (6.38). This is because of the fact that the electrons become heavy when moving in a periodic potential lattice, as is the case in a solid.

In metals, the electrons in the outermost shells are very weakly bound and are almost free to move about in the volume of the metal. The positive ions being heavy, are immobile and are confined to the lattice points. This is called the *free electron model*. The above calculation thus applies to the electrons in metals, which are fermions.

THERMIONIC EMISSION AND PHOTOELECTRIC EMISSION

The electrons in a metal are almost free to move within it. But for stability, the electrons must be contained within the metal by a suitable potential, such that even the electrons at the Fermi level require certain energy ϕ, called work function, to take it out of the metal, i.e., to infinity. Thus, $\phi = \varepsilon_0 - \varepsilon_F$, where ε_0 is the energy required to remove the zero energy electron. When metals are heated, they emit electrons. The phenomenon is called *thermionic emission*. We shall calculate the thermionic current density, i.e. the charge striking a unit normal area per unit time.

Let the x-axis of our coordinate system be normal to the metal surface. The rate at which electrons in the momentum range $\vec{p}$ and $\vec{p}+d\vec{p}$ strike unit area of the surface normal to x-axis is

$$v_x n(\vec{p})\, d\vec{p} = n(\vec{p})\, d\varepsilon_x\, dp_y\, dp_z$$

where v_x is the velocity in the x-direction, $\varepsilon_x = p_x^2/2m$ is the energy in the x-direction with $\varepsilon = \varepsilon_x + \varepsilon_y + \varepsilon_z$ and $n(\vec{p})$ is the number of electrons per unit volume of the phase space.

Since there are $2/h^3$ states per unit volume of phase space and electrons obey Fermi–Dirac statistics, we have

$$n(\vec{p}) = \frac{2}{h^3}\frac{1}{e^{(\varepsilon-\varepsilon_F)/kT}+1} \tag{6.39}$$

The current density j is the electronic charge e times the rate at which electrons having energy ε_x $(= p_x^2/2m) > \phi + \varepsilon_F$ strike unit area of the surface, i.e.,

$$j = \frac{2e}{h^3}\int_{-\infty}^{\infty}\int_{-\infty}^{\infty}\int_{\phi+\varepsilon_F}^{\infty}\frac{dp_z\,dp_y\,d\varepsilon_x}{e^{(\varepsilon-\varepsilon_F)/kT}+1}$$

$$= \frac{2ekT}{h^3}\int_{-\infty}^{\infty}\int_{-\infty}^{\infty}\ln\,(1+e^{-\theta})\,dp_y\,dp_z \tag{6.40}$$

where

$$\theta = \frac{1}{kT}\,(\phi+\varepsilon_y+\varepsilon_z) = \frac{1}{kT}\left(\phi+\frac{p_y^2+p_z^2}{2m}\right)$$

At ordinary temperatures $\theta \gg 1$, hence we may make a series expansion of the integrand and retain only the first term.

This gives

$$j = \frac{2ekT}{h^3}\,e^{-\phi/kT}\int_{-\infty}^{\infty}\int_{-\infty}^{\infty}e^{-(p_y^2+p_z^2)/2mkT}\,dp_y\,dp_z$$

$$= \frac{4\pi mek^2T^2}{h^3}\,e^{-\phi/kT}$$

or finally,

$$j = AT^2e^{-\phi/kT} \tag{6.41}$$

where

$$A = \frac{4\pi mek^2}{h^3} \tag{6.42}$$

This is the *Richardson–Dushman equation* for thermionic current density.

When light radiation of sufficient energy (photo energy is $h\nu$) falls on metals, the electrons may acquire enough energy to escape from the metal surface. The phenomenon is known as *photoelectric emission.*

If a photon of energy $h\nu$ strikes the metal and 1/2 mv^2 is the kinetic energy of the ejected photoelectron, then

$$h\nu - h\nu_0 = \frac{1}{2}\,mv^2 \tag{6.43}$$

where $h\nu_0$ is the minimum energy required for the ejection of a photoelectron and ν_0 is called threshold frequency. The energy to eject an electron from the lowest free electron state is $\varepsilon_F + \phi$. It is thus evident that $h\nu_0 = \phi$. For $\nu < \nu_0$, no photoemission occurs. This fact is utilized for the measurement of the work function. The expression for emission current can be obtained in a manner similar to that used for thermionic emission current. We quote the general expression given by Fowler

$$j = \frac{4\pi m e k^2 T^2}{h^3} \alpha F\left(\frac{h\nu - \phi}{kT}\right) \tag{6.44}$$

where α is the probability that an electron absorbs a photon to satisfy the condition for photoemission and

$$\begin{aligned} F(\delta) &\simeq e^{\delta} - \frac{e^{2\delta}}{2^2} + \frac{e^{3\delta}}{3^2} - \cdots \quad \text{for } \delta < 0 \\ &= \frac{\pi^2}{6} + \frac{\delta^2}{2} - \left(e^{-\delta} - \frac{e^{2\delta}}{2^2} + \frac{e^{3\delta}}{3^2} - \cdots\right) \quad \text{for } \delta > 0 \end{aligned} \tag{6.45}$$

SPIN PARAMAGNETISM

In most metals, the conduction electrons have a small temperature independent paramagnetic volume susceptibility of the order of 10^{-6}, that is in striking contrast with the Langevin formula (based on classical model) which gives a volume susceptibility of the order of 10^{-4} (at room temperature) and varies inversely with temperature. Pauli has shown that the application of Fermi–Dirac statistics corrects the theory as required.

The Hamiltonian of a non-relativistic free electron in the presence of a magnetic field $\vec{B}$ is

$$H = \frac{1}{2m}\left(\vec{p} + \frac{e}{c}\vec{A}\right)^2 - \hat{\mu}\cdot\vec{B} \tag{6.46}$$

where

$$\vec{B} = \text{curl } \vec{A} \tag{6.47}$$

$\vec{A}$ is the vector potential. The first term in Eq. (6.46) gives rise to diamagnetism and the second term to paramagnetism.

The intrinsic magnetic moment operator $\hat{\mu}$ is

$$\hat{\mu} = \mu\hat{\sigma} \tag{6.48}$$

with

$$\mu = \frac{e\hbar}{2mc} \tag{6.49}$$

and $\hat{\sigma}$ is spin operator and eigenvalues of $\hat{\sigma}\cdot\hat{B}$ are $\pm B$. Thus, the single particle energy levels are

$$\varepsilon_{p,s} = \frac{p^2}{2m} \pm \mu B \tag{6.50}$$

With the application of the external magnetic field $\vec{B}$, all electrons with a magnetic moment parallel to the field will suffer a shift in energy equal to $-\mu B$ and all the antiparallel ones will suffer a shift of μB. It should be noted that the shift $\mu B \ll \varepsilon_F$. Even for a strength of 10^5 gauss, $\mu B \simeq 10^{-3} eV$, much smaller than the Fermi energy of the order of a few electron volts.

We now evaluate the paramagnetic susceptibility per unit volume defined by

$$\chi = \frac{\partial M}{\partial B}, \ \chi > 0 \text{ paramagnetic}, \ \chi < 0 \text{ diamagnetic} \tag{6.51}$$

where M is the average induced magnetic moment per unit volume in the direction of the magnetic field, and it is assumed that the electron gas obeys Fermi–Dirac statistics.

We can write with the aid of Eq. (6.6)

$$g(\varepsilon \pm \mu B) = \frac{2\pi g V}{h^3}(2m)^{3/2}(\varepsilon \pm \mu B)^{1/2}$$

$$= \frac{2\pi g V}{h^3}(2m)^{3/2}\varepsilon^{1/2}\left(1 \pm \frac{\mu B}{\varepsilon}\right)^{1/2} \tag{6.52}$$

Since μB is small, we have

$$g(\varepsilon \pm \mu B) \simeq \frac{2\pi g V}{h^3}(2m)^{3/2}\varepsilon^{1/2}\left(1 \pm \frac{1}{2}\frac{\mu B}{\varepsilon}\right) \tag{6.53}$$

Hence we have for the net magnetization per unit volume,

$$M = \frac{\mu}{V}\int\left[\frac{1}{2}g(\varepsilon + \mu B) - \frac{1}{2}g(\varepsilon - \mu B)\right] f\, d\varepsilon$$

$$= \frac{2\pi g}{h^3}(2m)^{3/2}\frac{1}{2}\mu^2 B\int_0^\infty \frac{\varepsilon^{-1/2}\, d\varepsilon}{e^{(\varepsilon-\mu)/kT}+1} \tag{6.54}$$

If we set $\phi(\varepsilon) = \varepsilon^{-1/2}$, the above is the integral I of Eq. (6.18), and then use Eq. (6.26) for the value of the integral, retaining only the first term as the temperature is low, we get

$$M = \frac{2\pi g}{h^3}(2m)^{3/2}\mu^2 B\,\varepsilon_F^{1/2} \tag{6.55}$$

Use of Eq. (6.9) simplifies this to

$$M = \frac{3}{2}\frac{N}{V\varepsilon_F}\mu^2 B \tag{6.56}$$

The susceptibility per unit volume at low temperatures with $v = V/N$ is

$$\chi = \frac{\partial M}{\partial B} = \frac{3}{2}\frac{N\mu^2}{V\varepsilon_F} = \frac{3}{2}\frac{\mu^2}{vkT_F} \tag{6.57}$$

This is the *Pauli result*.

LANDAU DIAMAGNETISM

Landau first showed in 1930 that diamagnetism results from the quantization of orbits of the electrons. The magnetic properties of substances are due to electrons present in them. The atomic nuclei hardly make any contribution. The effects of an external magnetic field are: (a) the electrons, free or bound, move in quantized orbits normal to the magnetic field and (b) the spin of electrons tends to orient along the magnetic field. We have already seen the second to give rise to paramagnetism. To examine the first effect, we ignore the paramagnetism, i.e., we take a system of N spinless electrons. In the presence of the magnetic field $\vec{B}$, they move in circular orbits normal to $\vec{B}$ (Fig. 6.4) such that they satisfy the quantum conditions (old quantum theory, used for simplicity)

$$\oint \vec{p}.d\vec{r} = \left(j + \frac{1}{2}\right)h, \quad j = 0,1,2,\ldots \tag{6.58}$$

The Hamiltonian for a single electron is

$$H = \frac{1}{2m}\left(\vec{p} + \frac{e}{c}\vec{A}\right)^2 \tag{6.59}$$

and

$$\vec{B} = \text{curl } \vec{A} \tag{6.60}$$

The velocity of the electron has a constant magnitude and is tangential to the circular orbit (of radius a), and is given by

$$|\vec{v}| = \frac{ea}{mc}B \tag{6.61}$$

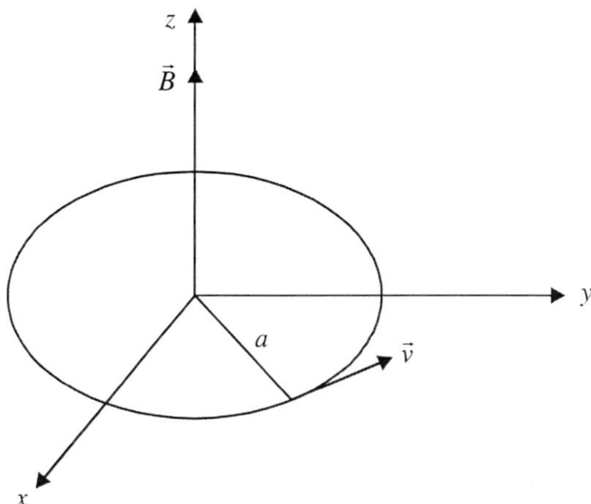

Figure 6.4 Electron orbit in a uniform magnetic field.

The canonical momentum $\vec{p}$ (not equal to $m\vec{v}$) is found from the Hamiltonian equation

$$\vec{v} = \nabla_p H = \frac{1}{m}\left(\vec{p} + \frac{e}{c}\vec{A}\right) \tag{6.62}$$

which gives

$$\vec{p} = m\vec{v} - \frac{e}{c}\vec{A} \tag{6.63}$$

Substituting Eq. (6.63) in the quantum condition Eq. (6.58), we get

$$\oint\left(m\vec{v} - \frac{e}{c}\vec{A}\right)\cdot d\vec{r} = \left(j + \frac{1}{2}\right)h \tag{6.64}$$

We have

$$\int m\vec{v}\cdot d\vec{r} = 2\pi a m v$$

and

$$\frac{e}{c}\oint \vec{A}\cdot d\vec{r} = \frac{e}{c}\oint \text{curl}\,\vec{A}\cdot d\vec{s} = \frac{e}{c}\,\pi a^2 B$$

This gives

$$2\pi a m v - \frac{e}{c}\pi a^2 B = \left(j + \frac{1}{2}\right)h \tag{6.65}$$

From this, we find that allowed orbits have radii satisfying the condition

$$a^2 = \frac{2c}{eB}\left(j + \frac{1}{2}\right)\hbar \tag{6.66}$$

The kinetic energy corresponding to the jth allowed orbit with the aid of Eqs. (6.61) and (6.66) is

$$\frac{1}{2}m|\vec{v}|^2 = \frac{eB}{mc}\left(j + \frac{1}{2}\right)\hbar \tag{6.67}$$

If p_z is the momentum associated with the motion along the z-direction, we have

$$p_z = \frac{2\pi\hbar}{V^{1/3}}l, \quad l = 0, \pm 1, \pm 2, \ldots \tag{6.68}$$

Thus, the allowed energies of an electron are the sum of $p_z^2/2m$ and that given by Eq. (6.67), i.e.

$$\varepsilon(p_z, j) = \frac{p_z^2}{2m} + \frac{eB}{mc}\left(j + \frac{1}{2}\right)\hbar \tag{6.69}$$

We thus see that the energy associated with a free electron in the absence of a magnetic field $(p_x^2 + p_y^2)/2m$ now corresponds to $(eB/mc)[j + (1/2)]\hbar$, which forms equispaced energy intervals of size $eB\hbar/mc$. All energy states originally in this interval assume the same energy in the presence of the magnetic field. Thus, the degeneracy g in $\varepsilon(p_z, j)$ is the number of quantum states satisfying the condition

$$\frac{p^2}{2m} = \frac{p_x^2 + p_y^2}{2m} \leq \frac{eB\hbar}{mc}$$

and is given by

$$g = \frac{2\pi V^{2/3}}{h^2} \int_0^{p \leq \sqrt{2eB\hbar/c}} p\,dp = \frac{V^{2/3} eB}{ch} \tag{6.70}$$

The presence of the magnetic field causes the above number of free electron energy level to degenerate as shown in the Fig. 6.5.

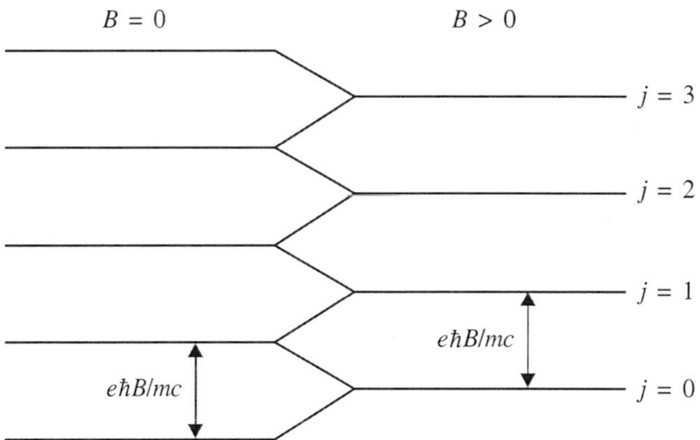

Figure 6.5 Energy spectra of electrons with and without magnetic field.

We incorporate this degeneracy to label the states of a single electron by a set of quantum numbers $p_z, j, \alpha \equiv \lambda$ and write

$$\varepsilon(p_z, j, \alpha) = \frac{p_z^2}{2m} + \frac{eB}{mc}\left(j + \frac{1}{2}\right)\hbar \tag{6.71}$$

where

$$p_z = \frac{2\pi\hbar}{V^{2/3}}\, l \tag{6.72}$$

where $l = 0, \pm1, \pm2, \ldots,\ j = 0, 1, 2, \ldots,\ \alpha = 1, 2, \ldots, g$.

Since electrons obey Fermi–Dirac statistics, we have

$$n_\lambda = 0, 1$$

$$\sum_\lambda n_\lambda = N \tag{6.73}$$

We can now write down the grand partition function

$$\Xi = \prod_\lambda (1 + z' e^{-\beta \varepsilon_\lambda}) \tag{6.74}$$

where $z' = e^{\mu/kT} = z\lambda^3$ is called activity and $\beta = 1/kT$.

This gives

$$\ln \Xi = \sum_{\alpha=1}^{g}\sum_{j=0}^{\infty}\sum_{p_z} \ln\left(1 + z' e^{-\beta\varepsilon_A(p_z, j, \alpha)}\right)$$

$$= g\,\frac{V^{1/3}}{h}\int_{-\infty}^{\infty} dp_z \sum_{j=0}^{\infty} \ln\left(1 + z' e^{-\beta\varepsilon(p_z, j, \alpha)}\right) \tag{6.75}$$

in which we summed over α and the summation over p_z has been replaced by integration.

At high temperatures $z' << 1$, we make a series expansion and retain only the first term to write

$$\ln \Xi \simeq \frac{z' g V^{1/3}}{h}\int_{-\infty}^{\infty} dp_z \sum_{j=0}^{\infty} e^{-\beta(e\hbar B/mc)[j + (1/2)]} e^{-\beta p_z^2/2m}$$

$$= \frac{z' V}{2\pi^2}\frac{eB}{\hbar^2 c}\sqrt{\frac{\pi m}{2\beta}}\, e^{-\beta e\hbar B/2mc}\left(1 + e^{-\beta e\hbar B/mc} + e^{-2\beta e\hbar B/mc} + \cdots\right)$$

$$= \frac{z' V}{2\pi^2}\frac{eB}{\hbar^2 c}\sqrt{\frac{\pi m}{2\beta}}\frac{e^{-x}}{1 - e^{-2x}} = \frac{z'}{2\pi^2}\frac{VeB}{\hbar^2 c}\sqrt{\frac{\pi m}{2\beta}}\frac{1}{2\sinh x} \tag{6.76}$$

where

$$x = \beta\frac{e\hbar}{2mc}B \tag{6.77}$$

and have used Eq. (6.70) for g.

We expand $\sinh x = x + (x^3/3!)$, and retain only first two terms to get

$$\ln \Xi \simeq \frac{z'}{4\pi^2}\frac{VeB}{\hbar^2 c}\sqrt{\frac{\pi m}{2\beta}}\;\frac{1}{x}\left(1 - \frac{x^2}{6}\right) = \frac{z' V}{\lambda^3}\left[1 - \frac{1}{6}\left(\frac{e\hbar}{2mc}\right)^2\left(\frac{B}{kT}\right)^2\right] \tag{6.78}$$

We can further write

$$\langle N\rangle \equiv N = z'\left(\frac{\partial \ln \Xi}{\partial z'}\right)_{B,V,T} = z\left(\frac{\partial \ln \Xi}{\partial z}\right)_{B,V,T} \tag{6.79}$$

to get

$$N \simeq \ln \Xi \tag{6.80}$$

and for $B \to 0$, we get

$$\frac{z'}{\lambda^3}(=z) = \frac{N}{V} = \rho = \frac{1}{v} \tag{6.81}$$

In the grand canonical ensemble, we have

$$M = \frac{1}{V}\left(-\frac{\partial H}{\partial B}\right) = \frac{kT}{V}\frac{\partial}{\partial B}\ln \Xi \tag{6.82}$$

This gives for magnetic susceptibility per unit volume, using Eqs. (6.78) and (6.82)

$$\chi = \frac{\partial M}{\partial B} = -\frac{1}{3}\frac{\rho}{kT}\left(\frac{e\hbar}{2mc}\right)^2 \tag{6.83}$$

This is in conformity with Curie law and $\rho/3\,(e\hbar/2mc)^2$ is the Curie constant.

THE EQUATION OF STATE AT HIGH DENSITY

For matter at high densities, when the volume per atom is less than the usual size of the atom, the matter is transformed into a highly compressed plasma of electrons and nuclei. The coulomb interaction energy per electron is Ze^2/a, where a is the mean distance between the electron and the nuclei ($a^3 \sim ZV/N$). The mean kinetic energy of the electrons is of the order of ε_F. When $\varepsilon_F >> Ze^2/a$, the electron gas can be taken as an ideal degenerate Fermi gas. The presence of the nuclei does not affect the thermodynamic properties of the gas. In the inequality $\varepsilon_F >> Ze^2/a$, if we substitute $a = (ZV/N)^{1/3}$ and use Eq. (6.9), we get

$$n_e\left(=\frac{N}{V}\right) \gg \left(\frac{m_e e^2}{\hbar^2}\right)^3 Z^2 \tag{6.84}$$

where n_e is the number density of electrons, m_e is electron mass and Z is the mean atomic number of the nuclei. The temperature corresponding to $(m_e e^2/\hbar^2)^3 Z^2$ is 1.26×10^6 K for $Z = 2$.

For n_e given by the inequality Eq. (6.84), the degeneracy temperature T_F becomes much higher than T and matter is a degenerate Fermi gas.

In terms of mass density $\rho(= n_e m')$, the inequality is

$$\rho \gg \left(\frac{m_e e^2}{\hbar^2}\right)^3 Z^2 m' \tag{6.85}$$

where m' is the mass per electron.

The right-hand side of Eq. (6.85) corresponds to 2×10^6 g/cc, and below this value the electron gas can be taken as non-relativistic and the pressure is given by Eq. (6.14).

$$P = \frac{(3\pi^2)^{2/3}}{5}\frac{\hbar^2}{m_e}\left(\frac{\rho}{m'}\right)^{5/3} \tag{6.86}$$

For densities comparable or higher than 2×10^6 g/cc, the electron gas becomes relativistic and pressure is given by Eq. (6.17)

$$P = \frac{1}{4}(3\pi^2)^{1/3} e\hbar \left(\frac{\rho}{m'}\right)^{4/3} \tag{6.87}$$

Further increase in density gives rise to nuclear reactions, consisting of capture of electrons by nuclei (protons of the nuclei change to neutrons) with emission of neutrinos. This results in the decrease of number of electrons but the equilibrium gives the electron density constant and hence the pressure does not change. At still higher densities, more nuclei will capture electrons and at about the mass density of the order of 10^{12} g/cc, the pressure due to neutrons (the neutrons being more numerous in comparison to the electrons) becomes dominant. In this region, the matter may be considered as a degenerate neutron Fermi gas (non-relativistic) and the equation of state is given by Eq. (6.14):

$$P = \frac{1}{5}(3\pi^2)^{2/3} \frac{\hbar^2}{m_n^{8/3}} \rho^{5/3} \tag{6.88}$$

where m_n is the neutron mass.

At densities $\rho >> 6 \times 10^{15}$ g/cc, the degenerate neutron gas becomes extreme relativistic and the equation of state is

$$P = \frac{1}{4}(3\pi^2)^{1/3} e\hbar \left(\frac{\rho}{m_n}\right)^{4/3} \tag{6.89}$$

However, it should be mentioned that at densities of the order of that of nuclear matter, the nuclear forces become significant, and the above expression is only qualitative. With our present state of knowledge, not much definite conclusions about matter at densities above the nuclear matter can be obtained.

WHITE DWARF STARS

The brightness of a star is proportional to its colour, i.e., the dominant wavelength, and the constant of proportionality is roughly the same for all stars. The plot of brightness against colour gives a linear strip called *main sequence*, in which most stars fall (Fig. 6.6). However, there are exceptions.

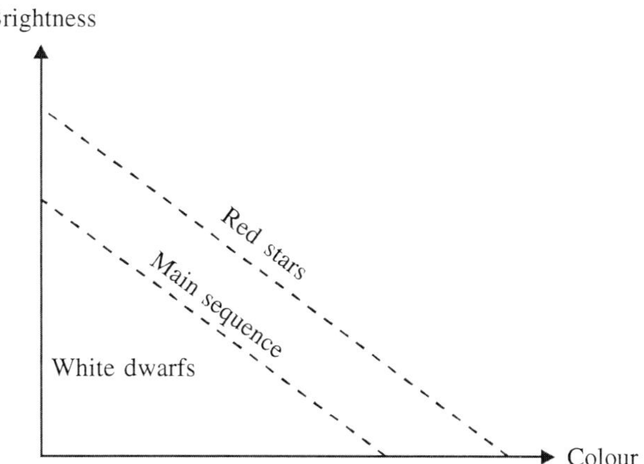

Figure 6.6 Brightness-colour diagram of stars (Russell–Hertzprung diagram).

There are huge red stars, abnormally bright for their colour and also white dwarf stars, abnormally faint for their white colour. The nearest white dwarf star to us is a companion of Sirius, eight-light years away. Their little brightness is derived from the release of gravitational energy through a slow contraction of the star. An idealized white dwarf star is a compressed plasma of electrons and nuclei (mostly helium), has mass of the order of that of the sun (2×10^{33} g = M_S) with mass density ~ 10^7 g/cc and the central temperature ~10^7 K. This corresponds to $n_e = 10^{30}$ per cc and the corresponding degeneracy temperature is 10^{11} K. Hence the electron gas is a highly degenerate cold Fermi gas.

We now examine the equilibrium of bodies of large mass held together by gravitational attraction. Actual stars radiate energy and are not in thermal equilibrium. However, we take an equilibrium body of large mass. The condition for the equilibrium of a body in a gravitational field is given by

$$\mu + m'\phi = \text{constant} \tag{6.90}$$

where μ is the chemical potential of the electron gas and ϕ is the gravitational potential. (The chemical potential of the nucleus being small is neglected.)

Using Newton gravitational formula for a spherically symmetric case, we have

$$\frac{1}{r^2}\frac{d}{dr}\left(r^2\frac{d\phi}{dr}\right) = 4\pi G\rho \tag{6.91}$$

where G is constant of gravitation.

From Eqs. (6.90) and (6.91), we get

$$\frac{1}{r^2}\frac{d}{dr}\left(r^2\frac{d\mu}{dr}\right) = -4\pi G\rho m' \tag{6.92}$$

For a non-relativistic degenerate electron gas, the chemical potential, using $\mu = \varepsilon_F$ and $\rho = n_e m'$ with $n_e = N/V$, can be written with the aid of Eq. (6.9) as

$$\mu = (3\pi^2)^{2/3}\frac{\hbar^2}{2m_e}\left(\frac{\rho}{m'}\right)^{2/3} \tag{6.93}$$

Substituting for ρ from the above into Eq. (6.92), we obtain

$$\frac{1}{r^2}\frac{d}{dr}\left(r^2\frac{d\mu}{dr}\right) = -K\mu^{3/2};\quad K = \frac{4}{3}\frac{(2m_c)^{3/2}}{\pi\hbar^3}m'^2 G \tag{6.94}$$

The constant K has the dimensions of $\text{cm}^{-2}\ \text{erg}^{-1/2}$. The solution of Eq. (6.94) should not have singularities at the origin, i.e. $\mu \to$ constant as $r \to 0$, Hence

$$\frac{d\mu}{dr} = 0 \qquad \text{for } r = 0 \tag{6.95}$$

This gives

$$\frac{d\mu}{dr} = -\frac{K}{r^2}\int_0^r r^2\mu^{3/2} \tag{6.96}$$

Since $1/K^2R^4$ has the dimensions of energy, and r/R being dimensionless, we can write $\mu(r)$ in the form:

$$\mu(r) = \frac{1}{K^2R^4}f\left(\frac{r}{R}\right) \tag{6.97}$$

and for the density ($\rho \propto \mu^{3/2}$)

$$\rho(r) = \frac{\text{constant}}{R^6} F\left(\frac{r}{R}\right) \tag{6.98}$$

where f and F are some functions of the dimensionless quantity $r/R = \xi$ (say).

This gives

$$\rho \propto \frac{1}{R^6} \text{ and } M \propto \frac{1}{R^3} \tag{6.99}$$

The total kinetic energy of a non-relativistic degenerate electron gas from Eq. (6.13) is proportional to $N(N/V)^{2/3}$, i.e., proportional to $M^{5/3}R^{-2}$ and the gravitational energy is proportional to $-M^2R^{-1}$. The total of these two has a minimum for all M and at minimum, $R \propto M^{-1/3}$. Thus, a gravitating sphere of non-relativistic degenerate Fermi gas can be in equilibrium for any value of total mass.

Substitution of Eq. (6.97) into Eq. (6.94) yields

$$\frac{1}{\xi^2}\frac{d}{d\xi}\left(\xi^2 \frac{df}{d\xi}\right) = -f^{3/2} \tag{6.100}$$

with boundary conditions

$$f'(0) = 0, \; f(1) = 0$$

Equation (6.100) cannot be solved analytically. Numerical solution gives

$$f(0) = 178.2, \quad f'(1) = -132.4$$

Multiplying both sides of Eq. (6.91) by $r^2\,dr$ and integrating from 0 to R

$$GM = R^2\left(\frac{d\phi}{dr}\right)_{r=R} = -\left(\frac{R^2}{m'}\right)\left(\frac{d\mu}{dr}\right)_{r=R} = -\frac{f'(1)}{m'K^2R^3}$$

Substituting the value of $f'(1)$ and K from Eq. (6.94), we obtain

$$MR^3 = 91.9\left(\frac{\hbar^6}{G^3m_e^3m'^5}\right) = 2.2\times10^{13}\left(\frac{m_n}{m'}\right)^5 M_s\,\text{km}^3 \tag{6.101}$$

Let us now turn to the equilibrium of a sphere consisting of extreme relativistic degenerate electron gas, as is the case in white dwarf stars. The total kinetic energy of an extreme relativistic electron gas from Eq. (6.16) is proportional to $N(N/V)^{1/3}$, i.e., to $M^{4/3}/R$ and the resulting pressure $P = -(\partial E/\partial V)_s$ acts outwards. For equilibrium, the sum of kinetic energy and the gravitation energy $-(M^2/R)$ (attractive) must vanish. This qualitative reasoning is confirmed by an exact quantitative analysis. The chemical potential, using $\mu = \varepsilon_F$, $\rho = n_em'$ and Eq. (6.16), is written as

$$\mu = (3\pi^2)^{1/3}\, e\hbar\left(\frac{\rho}{m'}\right)^{1/3} \tag{6.102}$$

and in place of Eq. (6.94), we now have

$$\frac{1}{r^2}\frac{d}{dr}\left(r^2\frac{d\mu}{dr}\right) = -K'\mu^3; \quad K' = \frac{4}{3}\frac{1}{\pi c^3\hbar^3}m'^2G \tag{6.103}$$

K' has dimensions of cm^{-2} erg^{-2} and hence we can write the chemical potential in the form

$$\mu(r) = \frac{1}{R\sqrt{K'}} f\left(\frac{r}{R}\right) \tag{6.104}$$

and the density ($\rho \propto \mu^3$) in the form

$$\rho(r) = \frac{\text{constant}}{R^3} F\left(\frac{r}{R}\right) \tag{6.105}$$

where f and F are some functions of r/R $(=\xi)$.

We immediately find that the total mass $M \propto R^3\rho$ and is independent of R, i.e.,

$$M = \text{constant} = M_0 \text{ (say)}$$

where M_0 is the value of the mass for which equilibrium is possible. For $M > M_0$, the gravitational attraction dominates and the body will contract indefinitely. For $M < M_0$, the outward pressure dominates and the body will keep expanding. To obtain the value of M_0, we substitute the value of μ from Eq. (6.104) into Eq. (6.103), to get

$$\frac{1}{\xi^2}\frac{d}{d\xi}\left(\xi^2\frac{df}{d\xi}\right) = -f^3 \tag{6.106}$$

with the boundary condition $f'(0) = 0$ and $f(1) = 0$ as before. The analytical solution of Eq. (6.106) not being possible, it is solved numerically. The solution gives

$$f(0) = 6.897, \quad f'(1) = -2.018$$

The total mass M_0 is obtained by multiplying Eq. (6.91) by $r^2\,dr$ and integrating from 0 to R:

$$GM_0 = R^2\left(\frac{d\phi}{dr}\right)_{r=R} = -\frac{f'(1)}{m'\sqrt{K'}}$$

whence

$$M_0 = \frac{3.1}{m'^2}\left(\frac{c\hbar}{G}\right)^{3/2} = 5.8\left(\frac{m_n}{m'}\right)^2 M_s \tag{6.107}$$

If we take $m' = 2m_n$, we get

$$M_0 = 1.45M_s \tag{6.108}$$

The result was first obtained in 1931 by S. Chandrasekhar (*Stellar Structure*, Dover, New York, 1957, Chap. XI) and is known as *Chandrasekhar limit*. This result is supported by astronomical observations.

NEUTRON STARS

It has already been stated that at $\rho \sim 3 \times 10^{11}$ g/cc (corresponding $P \sim 10^{24}$ bar), the number of neutrons is much larger than that of the electrons and the matter may be treated essentially as a degenerate neutron Fermi gas forming a neutron star.

To examine the equilibrium of a neutron sphere, we again start with the equilibrium condition

$$\mu + m_n\phi = \text{constant} \tag{6.109}$$

where m_n is the neutron mass and ϕ the gravitational potential.

The pressure at the boundary of the body must vanish; hence the outer layer of the body is at low pressure and density. It consists of electrons and nuclei. The width of outer shell is comparable with the radius of the dense inner neutron core, but its density being very low, the total mass of the body may be taken as the mass of the inner core. Let R and R' be the radii of the inner core and the outer shell respectively. The difference between chemical potentials at these points is equal to the difference between the rest energy of a neutral atom (a nucleus and Z electrons) per unit atomic weight and the rest energy of the neutron. If this difference between these two rest energies is Δ, then

$$Gm_nM\left(\frac{1}{R} - \frac{1}{R'}\right) = \Delta$$

This gives

$$\frac{Gm_nM}{R} > \Delta \tag{6.110}$$

As the neutron gas is non-relativistic, we have from Eq. (6.101) with $m' = m_n$, $m_e = m_n$

$$MR^3 = 91.9\left(\frac{\hbar^6}{G^3m_n^8}\right) = 3.6 \times 10^3 M_s \tag{6.111}$$

Substituting the value of R obtained from the above equation in Eq. (6.110), we get

$$M > \sim 0.2M_s \tag{6.112}$$

and the corresponding radius $R < 26$ km. Beyond this lower limit of mass, the neutron state of the body cannot be stable.

To find the upper limit on the mass, we have to go to the relativistic case and use Eq. (6.107) with $m' = m_n$ to get

$$M < 6M_s$$

But this result is not true, because in a relativistic neutron gas, the kinetic energy of the particle is of the same order as or even greater than the rest energy and the gravitational potential $\phi \sim c^2$ (Volkoff, G.M., Phys. Rev. **55**, 374, 1939). In these circumstances, the Newtonian gravitation theory is not valid, and recourse must be taken to the general theory of relativity. The result obtained is

$$M < 0.76M_s \tag{6.113}$$

and the corresponding radius $R_{\min}$ is 9.4 km.

A spherical body of mass greater than above will tend to contract indefinitely, i.e., there will be an unrestrained gravitational collapse. They are known as *black holes.*

EXERCISES

1. Derive the Fermi–Dirac distribution law using grand canonical ensemble. Show that $PV = (2/3)\,E$ for an ideal monoatomic Fermi–Dirac gas.

Hint: $\ln \Xi \Big|_{B.E.}^{F.D.} = -\sum_i \frac{\Omega_i}{kT} = \pm \sum_i \ln\,[1 \pm e^{\beta(\mu-\varepsilon_i)}]$

The upper sign is for a Fermi–Dirac gas and the lower sign for a Bose–Einstein gas

$$g(\varepsilon)d\varepsilon = \frac{2\pi gV}{h^2}(2m)^{3/2}\,\varepsilon^{1/2}d\varepsilon$$

$$E\Big|_{B.E.}^{F.D.} = \frac{2\pi gV}{h^3}(2m)^{3/2}\int_0^\infty \frac{\varepsilon^{3/2}d\varepsilon}{e^{\beta(\varepsilon-\mu)} \pm 1}$$

$$PV\Big|_{B.E.}^{F.D.} = kT\,\ln \Xi\Big|_{B.E}^{F.D.} = \pm\, kT \sum_i \ln\,[1 \pm e^{\beta(\mu-\varepsilon_i)}]$$

Replacing the summation by integration, we have

$$PV\Big|_{B.E.}^{F.D.} = \pm\, kT\,\frac{2\pi gV}{h^2}(2m)^{3/2}\int_0^\infty d\varepsilon\,\varepsilon^{1/2}\ln\,[1 \pm e^{\beta(\mu-\varepsilon)}]$$

Integrating by parts, we get

$$PV\Big|_{B.E.}^{F.D.} = \frac{2\pi gV}{h^2}(2m)^{3/2}\,\frac{2}{3}\int_0^\infty \frac{\varepsilon^{3/2}\,d\varepsilon}{e^{\beta(\varepsilon-\mu)} \pm 1} = \frac{2}{3}E$$

2. Show that the quantal correction term in the equation of state of an ideal monoatomic Fermi–Dirac gas is positive.

Hint: $\frac{PV}{KT}\Big|_{B.E.}^{F.D.} = \ln \Xi = \pm \sum_i \ln\,[1 \pm e^{\beta(\mu-\varepsilon_i)}]$

$$N\Big|_{B.E.}^{F.D.} = \sum_i \langle n_i \rangle = \sum_i [e^{\beta(\mu-\varepsilon_i)} \pm 1]^{-1}$$

This gives

$$P - \frac{NkT}{V} = \mp \frac{kT}{V}\sum_i [\ln\,(1 \mp \langle n_i \rangle) \pm \langle n_i \rangle]$$

For a Fermi–Dirac gas, we have $0 \le \langle n_i \rangle \le 1$; the expression under the bracket is negative and hence the right-hand side is positive. For a Bose–Einstein gas, we have $\langle n_i \rangle \ge 0$; the expression under the bracket is negative and hence the right-hand side is negative. Thus, the quantal correction to the pressure is positive for a Fermi–Dirac gas (additional repulsion due to the Pauli principle) and negative for a Bose–Einstein gas (as though some attraction exists between the particles).

3. Find the quantal correction to the equation of state of an ideal quantal gas.

Hint: $$\ln \Xi \Bigg|_{B.E.}^{F.D.} = \pm \sum_i \ln[1 \pm e^{\beta(\mu-\varepsilon_i)}] = \pm \sum \ln (1 \pm z' e^{-\varepsilon_i/kT})$$

$$= z' q(T) \mp \frac{1}{2} z'^2 q\left(\frac{T}{2}\right) + \frac{1}{3} z'^3 q\left(\frac{T}{3}\right)$$

$q(T)$ is single particle partition function at temperature T given by Eq. (3.22). From Eq. (4.28), we have

$$N \Bigg|_{B.E.}^{F.D.} = z' q(T) \mp z'^2 q\left(\frac{T}{2}\right) + z'^3 q\left(\frac{T}{3}\right) + \cdots$$

This gives

$$\frac{PV}{NkT} = 1 \pm \frac{1}{2} z' \frac{q(T/2)}{q(T)} - z'^2 \left[\frac{2}{3} \frac{q(T/3)}{q(T)} - \frac{1}{2} \frac{q^2(T/2)}{q^2(T)}\right] + \cdots$$

or

$$PV = NkT \left\{1 \pm \frac{1}{2} z' \frac{q(T/2)}{q(T)} - z'^2 \left[\frac{2}{3} \cdot \frac{q(T/3)}{q(T)} - \frac{1}{2} \frac{q^2(T/2)}{q^2(T)}\right] + \cdots\right\}$$

4. What is Fermi energy and what is its thermodynamic meaning? Show that a Fermi gas possesses energy and exerts pressure even at absolute zero.

5. Calculate the energy of a degenerate Fermi gas at very low temperatures and hence obtain its specific heat. How does it compare with experimental observations?

6. Discuss Pauli theory of paramagnetism. How does it differ from Langevin theory?

7. Discuss Landau theory of diamagnetism for an ideal Fermi gas.

8. Discuss the existence of white dwarf stars and find the limit for its maximum mass.

9. How are the thermionic and the photoelectric current densities obtained, assuming electrons obey Fermi–Dirac statistics?

10. Find C_v for $T < T_F$ for liquid He^3 assuming it to be a Fermi–Dirac gas of specific volume $V/N = 46.2$ Å^3/atom and compare it with the observed value of $2.89NkT$ for $T = 0.1$ K by Anderson et al.

Hint: $C_v = \dfrac{\pi^2}{2} \dfrac{Nk^2 T}{\varepsilon_F}$ and $\varepsilon_F = \dfrac{h^2}{2m} \left(\dfrac{3}{4\pi g} \dfrac{N}{V}\right)^{2/3}$

7 Ideal Bose Gas

THE BOSE–EINSTEIN DISTRIBUTION

We now turn to the study of statistics of a perfect gas of particles, which are described by symmetric wave functions. Such a statistics is called *Bose–Einstein statistics*.

For bosons, the occupation number of quantum states is not restricted at all and can take any value like 0, 1, 2, ... To obtain the distribution, we proceed as in Eq. (6.1) to write for the grand potential from Eq. (4.24):

$$\Omega_i = -kT \ln \sum_{n_i=0}^{\infty} (e^{(\mu-\varepsilon_i)/kT})^{n_i} \tag{7.1}$$

The summation on the right-hand side is a geometric series, which converges only if $e^{(\mu-\varepsilon_i)/kT} < 1$. Since Eq. (7.1) must hold for all values of ε_i, including $\varepsilon_i = 0$, we must have $e^{\mu/kT} < 1$, i.e.,

$$\mu < 0 \tag{7.2}$$

Thus in Bose statistics, chemical potential must always be negative. Recall that in the Boltzmann statistics, the chemical potential is negative but $|\mu|$ is large and in Fermi statistics, the chemical potential is either negative or positive but $|\mu|$ is small.

Summing the geometric series, we obtain

$$\begin{aligned}\Omega_i &= -kT \ln (1 - e^{(\mu-\varepsilon_i)/kT})^{-1} \\ &= kT \ln (1 - e^{(\mu-\varepsilon_i)/kT})\end{aligned} \tag{7.3}$$

Differentiating the above with respect to μ with V and T constant and using Eq. (4.28), we get

$$\langle n_i \rangle = \frac{1}{e^{(\varepsilon_i-\mu)/kT} - 1} \tag{7.4}$$

Dropping the subscript, we write

$$f = \frac{1}{e^{(\varepsilon-\mu)/kT} - 1} \tag{7.5}$$

This is the distribution function for an ideal Bose gas. It should be noted that the Fermi distribution and the Bose distribution tend to Boltzmann distribution if $e^{(\mu-\varepsilon_i)/kT} \ll 1$.

The Bose distribution is normalized by the condition

$$\sum_i \frac{1}{e^{(\varepsilon_i-\mu)/kT} - 1} = N \tag{7.6}$$

where N is the total number of particles in the gas.

For such a gas, the number of particles dN_ε in the energy range ε and $\varepsilon + d\varepsilon$ is

$$dN_\varepsilon = g(\varepsilon) f\, d\varepsilon = \frac{2\pi g V}{h^3} (2m)^{3/2} \frac{\varepsilon^{1/2} d\varepsilon}{e^{(\varepsilon-\mu)/kT} - 1} \tag{7.7}$$

where $g = 2s + 1$.

Integration over the energy gives the total number of particles N:

$$N = \frac{2\pi g V}{h^3} (2m)^{3/2} \int_0^\infty \frac{\varepsilon^{1/2}\, d\varepsilon}{e^{(\varepsilon-\mu)/kT} - 1} \tag{7.8}$$

This also determines μ as function of N/V and T.

The integral of the kind occurring in Eq. (7.8) are evaluated by a class of functions (Robinson, J.E., *Phys. Rev.* **83**, 678, 1951) defined by

$$g_s(z') = \frac{1}{\Gamma(s)} \int_0^\infty \frac{dx\, x^{s-1}}{(1/z')e^x - 1} = \sum_{k=1}^\infty \frac{z'^k}{k^s} \tag{7.9}$$

Proof:

$$g_s(z') = \frac{1}{\Gamma(s)} \int_0^\infty \frac{dx\, x^{s-1}\, z' e^{-x}}{(1 - z'e^{-x})} = \frac{1}{\Gamma(s)} \int_0^\infty dx\, x^{s-1}\, z' e^{-x}\, (1 + z' e^{-x} + z'^2 e^{-2x} + \cdots)$$

$$= \frac{1}{\Gamma(s)} \int_0^\infty dx\, x^{s-1} \sum_{k=1}^\infty z'^k e^{-kx} = \frac{1}{\Gamma(s)} \sum_{k=1}^\infty z'^k \int_0^\infty dx\, x^{s-1}\, e^{-kx}$$

$$= \frac{1}{\Gamma(s)} \sum_{k=1}^\infty \frac{z'^k}{k^s} \int_0^\infty dt\, t^{s-1}\, e^{-t} \qquad \text{where } t = kx$$

$$= \sum_{k=1}^\infty \frac{z'^k}{k^s}$$

If we identify $\varepsilon/kT = x$ with $e^{\mu/kT} = z'$, we have

$$N = gV \left(\frac{2\pi m kT}{h^2} \right)^{3/2} g_{3/2}(z') \tag{7.10}$$

DEGENERATE BOSE GAS

At low temperatures, the properties of a Bose gas are completely different from that of a Fermi gas. This is obvious from the fact that at $T = 0$ K, a Bose gas is in the state of lowest energy $E = 0$ (every particle in the quantum state $\varepsilon = 0$) while a Fermi gas has a non-zero energy.

For a given density $\rho = N/V$, if the temperature of the gas is lowered, the chemical potential given by Eq. (7.8) will rise. Since μ is negative, it will reach the value $\mu = 0$ at a temperature T_0 determined by Eq. (7.10), with $z' = 1$:

$$\frac{N}{V} = g \left(\frac{2\pi mkT_0}{h^2} \right)^{3/2} g_{3/2}(z'=1) \tag{7.11}$$

We see from Eq. (7.9) that $g_{3/2}$ $(z' = 1) = \zeta(3/2)$. This gives

$$T_0 = \frac{h^2}{2\pi mk} \left(\frac{N/V}{2.612\, g} \right)^{2/3} \tag{7.12}$$

For $T < T_0$, Eq. (7.8) has no negative solutions but in Bose statistics, μ must always be negative. This contradiction arises because of the improper replacement of the sum in Eq. (7.6) by an integration in Eq. (7.8). The first term in the sum $\varepsilon_i = 0$ is multiplied by $\sqrt{\varepsilon_i} = 0$ and disappears from the sum. Mathematically, we find that at $T = 0$ K, as $\mu \to 0$, the sum of all the terms in Eq. (7.6) except the first term tends to a finite limit as given by the integral in Eq. (7.8). The first term tends to infinity. We can make the first term finite by letting μ tend to some small finite value and not zero.

For $T < T_0$, μ is very small (very close to zero). This gives $e^{\mu/kT} \to 1$ and we have from Eq. (7.8)

$$\begin{aligned} N_{\varepsilon > 0} &= \frac{2\pi gV}{h^3} (2m)^{3/2} \int_o^\infty \frac{\varepsilon^{1/2} d\varepsilon}{e^{\varepsilon/kT} - 1} \\ &= gV \left(\frac{2\pi mkT}{h^2} \right)^{3/2} g_{3/2}(z'=1) \\ &= N \left(\frac{T}{T_0} \right)^{3/2} \end{aligned} \tag{7.13}$$

and the remaining number of particles with $\varepsilon = 0$ is

$$N_{\varepsilon=0} = N \left[1 - \left(\frac{T}{T_0} \right)^{3/2} \right] \tag{7.14}$$

which gives the number of particles in the lowest energy state $\varepsilon = 0$. The effect of concentrating the particles in the lowest energy state ($\varepsilon = 0$) is called *Bose–Einstein condensation*. At $T = T_0$, we see from Eq. (7.13) that $N_{\varepsilon>0} \simeq N$. As temperature is lowered, particles begin to concentrate in the state $\varepsilon = 0$, and the gas becomes degenerate. For this reason, T_0 is also called the degeneracy temperature. It should be kept in mind that the condensation is only in the momentum space and no condensation of the gas actually occurs.

The energy can be calculated as below

$$E = \int \varepsilon \, dN_\varepsilon = \frac{2\pi g V}{h^3}(2m)^{3/2} \int_0^\infty \frac{\varepsilon^{3/2}\, d\varepsilon}{e^{(\varepsilon-\mu)/kT} - 1} \tag{7.15}$$

$$= \frac{2\pi g V}{h^3} kT(2mkT)^{3/2} \int_0^\infty \frac{x^{3/2}\, dx}{\frac{1}{z'}e^x - 1}$$

where $x = \varepsilon/kT$ and $z' = e^{\mu/kT}$.

The integral on the right-hand side is same as that in Eq. (7.9), with $s = 5/2$. This gives

$$E = \frac{2\pi g VkT}{h^3}(2mkT)^{3/2}\,\Gamma\left(\frac{5}{2}\right) g_{5/2}(z')$$

$$= \frac{3}{2} kTgV \left(\frac{2\pi m kT}{h^2}\right)^{3/2} g_{5/2}(z') \tag{7.16}$$

We now distinguish the two cases for $T < T_0$ and $T > T_0$ by writing them as E_- and E_+.

For $T < T_0$, $z' \to 1$, and we have from Eq. (7.16)

$$E_- = \frac{3}{2} kTgV \left(\frac{2\pi m kT}{h^2}\right)^{3/2} g_{5/2}(z' = 1)$$

We substitute for gV from Eq. (7.11) to get

$$E_- = \frac{3}{2} NkT \left(\frac{T}{T_0}\right)^{3/2} \frac{\zeta(5/2)}{\zeta(3/2)}$$

$$= \frac{3}{2} NkT \left(\frac{T}{T_0}\right)^{3/2} \times 0.51 = 0.77 NkT \left(\frac{T}{T_0}\right)^{3/2} \tag{7.17}$$

For $T > T_0$, $z' < 1$ and $N_{\varepsilon>0} \simeq N$, we have from Eq. (7.16)

$$E_+ = \frac{3}{2} kTgV \left(\frac{2\pi m kT}{h^2}\right)^{3/2} g_{5/2}(z')$$

We substitute for gV from Eq. (7.10) to get

$$E_+ = \frac{3}{2} NkT \frac{g_{5/2}(z')}{g_{3/2}(z')}$$

The ratio $g_{5/2}(z')/g_{3/2}(z')$ can be obtained by assuming a power series expansion of $g_{5/2}(z')$ in terms of $g_{3/2}(z')$ and finding the coefficients using Taylor theorem. The result is

$$\frac{g_{5/2}(z')}{g_{3/2}(z')} = \left[1-0.177 g_{3/2}(z') - 0.003\, g_{3/2}(z')^2 - \cdots\right]$$

Combining Eqs. (7.10) and (7.11), we write

$$g_{3/2}(z') = \left(\frac{T_0}{T}\right)^{3/2} g_{3/2}(z'=1) = \left(\frac{T_0}{T}\right)^{3/2} \zeta\left(\frac{3}{2}\right)$$

to finally get

$$E_+ = \frac{3}{2} NkT\left[1 - 0.462\left(\frac{T_0}{T}\right)^{3/2} - 0.23\left(\frac{T_0}{T}\right)^3 - \cdots\right] \tag{7.18}$$

This gives for specific heat

$$C_{V-} = \frac{15}{4} \times 0.51 Nk\left(\frac{T}{T_0}\right)^3 = 1.926 Nk\left(\frac{T}{T_0}\right)^{3/2} \tag{7.19}$$

$$C_{V+} = \frac{3}{2} Nk\left[1 + 0.231\left(\frac{T_0}{T}\right)^{3/2} + 0.045\left(\frac{T_0}{T}\right)^3 + \cdots\right] \tag{7.20}$$

We see that the specific heat varies as $T^{3/2}$ nears absolute zero. At $T = T_0$, we find that $C_{V-} = C_{V+}$ but their slopes are different. This means the specific heat is continuous at $T = T_0$, but a graph of C_V/Nk against T will show a cusp at $T = T_0$ as shown in Fig. 7.1.

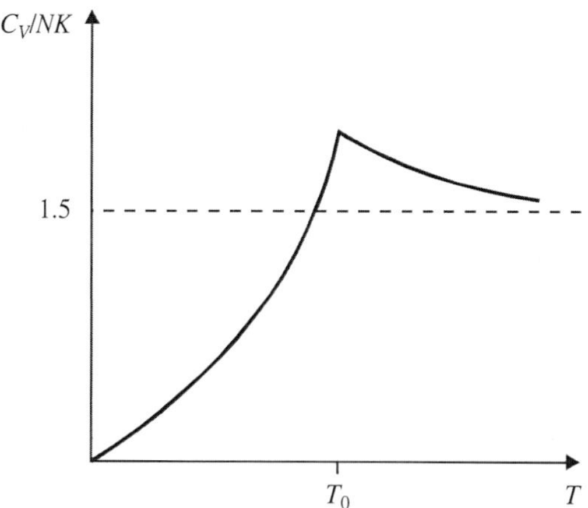

Figure 7.1 Variation of specific heat of an ideal Bose gas.

The only Bose system known to exist at low temperatures is liquid He^4, which we will discuss later separately.

BLACKBODY RADIATION

Electromagnetic radiation in thermal equilibrium is called *blackbody radiation*. Planck in 1900 proposed that the radiation is emitted or absorbed in packets of energy $\varepsilon = h\nu$. Einstein, in 1905, showed that the radiation also travels as packets or quanta of energy called photons (spin 1). The photons do not interact with one another and hence form an ideal Bose gas.

For radiation in a material medium, the interaction between radiation and matter must be small for the gas to be considered as an ideal gas. This condition is satisfied for gases at all frequencies, except near the absorption lines of the medium. For a medium of high density, it is true only at high temperatures.

Since there is no interaction between the photons, the presence of the material medium is necessary to enable thermal equilibrium to be set up. The thermal equilibrium is brought about by the absorption and the emission of photons by the matter present. Thus, the number of photons N is not a fixed quantity and has to be determined from the condition of thermal equilibrium. In thermal equilibrium, the free energy is minimum for a given temperature and volume. Thus, the necessary condition is

$$\left(\frac{\partial A}{\partial N}\right)_{V,T} = 0 \tag{7.21}$$

Since

$$\left(\frac{\partial A}{\partial N}\right)_{V,T} = \mu$$

We obtain the result that the chemical potential of the photon gas is zero, i.e.,

$$\mu = 0 \tag{7.22}$$

For photons $E = h\nu$, $p = h\nu/c$, we have for the number of quantum states between frequencies ν and $\nu + d\nu$ (corresponding to momentum between p and $p + dp$)

$$\frac{4\pi V}{c^3}\nu^2 d\nu$$

in which we have moved from a discrete to a continuous distribution of frequencies assuming V to be large. We multiply the above by 2, as there are two independent directions of polarization, to obtain the number of quantum states between frequencies ν and $\nu + d\nu$ to be

$$g(\nu)d\nu = \frac{8\pi V}{c^3}\nu^2 d\nu \tag{7.23}$$

This gives for the number of photons in the frequency range ν and $\nu + d\nu$, obeying the Bose distribution

$$dN_\nu = \frac{8\pi V}{c^3}\frac{\nu^2 d\nu}{e^{h\nu/kT} - 1} \tag{7.24}$$

Multiplying the above by $h\nu$, we obtain the radiation energy in this region of the spectrum

$$dE_\nu = \frac{8\pi Vh}{c^3} \frac{\nu^3 d\nu}{e^{h\nu/kT} - 1} \tag{7.25}$$

or in terms of the wavelength λ

$$dE_\lambda = \frac{16\pi^2 c\hbar V}{\lambda^5} \frac{d\lambda}{e^{2\pi\hbar c/\lambda kT} - 1} \tag{7.26}$$

The above expression for the spectral energy distribution of blackbody radiation is called *Planck formula*, which laid the foundation of the quantum theory.

For low frequencies $h\nu << kT$, we expand $e^{h\nu/kT} \simeq 1 + h\nu/kT$ to get

$$dE_\nu = \frac{8\pi V}{c^3} kT\nu^2 \, d\nu \tag{7.27}$$

which is the *Rayleigh–Jeans law*.

For high frequencies $h\nu >> kT$, unity in the denominator is neglected to yield

$$dE_\lambda = 16\pi^2 c\hbar V \lambda^{-5} e^{-2\pi\hbar c/\lambda kT} \, d\lambda \tag{7.28}$$

which is the *Wien formula*.

The total energy density u is obtained by integrating Eq. (7.25), with $x = h\nu/kT$:

$$u = \frac{E}{V} = \frac{8\pi h}{c^3} \int_0^\infty \frac{\nu^3 \, d\nu}{e^{h\nu/kT} - 1} = \frac{8\pi h}{c^3} \left(\frac{kT}{h}\right)^4 \int_0^\infty \frac{x^3 \, dx}{e^{x-1}} \tag{7.29}$$

which is evaluated with the aid of Eq. (7.9) to yield

$$u = \frac{E}{V} = \frac{8\pi}{c^3 h^3} (kT)^4 \, \Gamma(4) \, \zeta(4) = \frac{8\pi k^4}{c^3 h^3} \frac{\pi^4}{15} T^4$$

$$= \frac{4\sigma}{c} T^4 \tag{7.30}$$

which is called the *Stefan–Boltzmann law* and where

$$\sigma = \frac{\pi^2 k^4}{60 c^2 \hbar^3} = 5.67 \times 10^{-8} \text{ W m}^{-2} \text{ K}^{-4}$$

is called the *Stefan–Boltzmann constant*.

The free energy $A = \mu N + \Omega = \Omega(\mu = 0) = \sum_i \Omega_i$, is given by

$$A = \frac{8\pi V}{c^3} kT \int_0^\infty d\nu \, \nu^2 \ln(1 - e^{-h\nu/kT}) = -\frac{4\sigma}{3c} VT^4 \tag{7.31}$$

and the pressure of the radiation

$$P = -\left(\frac{\partial A}{\partial V}\right)_T = \frac{4\sigma}{3c} T^4 = \frac{u}{3} \tag{7.32}$$

The entropy is

$$S = -\left(\frac{\partial A}{\partial T}\right)_V = \frac{16\sigma}{3c} VT^3 \tag{7.33}$$

and the specific heat of radiation is

$$C_V = \left(\frac{\partial E}{\partial T}\right)_V = \frac{16\sigma}{c} VT^3 \tag{7.34}$$

PHONONS IN SOLIDS

The study of solids forms, an important branch of physics, called *solid-state physics*. We concern ourselves only with the thermal behaviour of solids here. The earliest correlations in solids was the *Dulong and Petit law* (1819) that the atomic heat capacity (product of mass number (atomic weight) and specific heat per unit mass) was about 6 cal/g atom degree for a large number of elements. It was explained by Boltzmann (1871) using equipartition theorem. The atoms in solids are restricted to lattice sites but can vibrate about their mean positions. The Hamiltonian of a three-dimensional oscillator is given by

$$H = \frac{\vec{p}^2}{2m} + \frac{1}{2} m\omega^2 \vec{r}^2$$

and has six degrees of freedom. Thus, the mean energy of the oscillator is $3NkT$ (N is the number of atoms or molecules) and the specific heat is $3Nk$, which is the Dulong and Petit law. However, it was observed towards the end of nineteenth century that solids like beryllium, diamond, etc. had low values of atomic heat capacities and approached the Dulong and Petit behaviour only at high temperatures. Also, experiments have shown that the specific heat of most materials decreased as temperature is lowered. Classical statistical mechanics fails to explain this behaviour.

Einstein (1907) applied the quantum ideas to the vibration of atoms and explained the freezing and the thawing of degrees of freedom. The energy levels of an oscillator of frequency ν is

$$\varepsilon_n = \left(n + \frac{1}{2}\right) h\nu$$

The partition function is then

$$Q = \sum_{n=0}^{\infty} e^{-\varepsilon_n/kT} = e^{-h\nu/2kT} \sum_{n=1}^{\infty} e^{-nh\nu/kT} = \frac{e^{-h\nu/2kT}}{1 - e^{h\nu/kT}} \tag{7.35}$$

and the energy is

$$\varepsilon = -\frac{\partial \ln Q}{\partial \beta} = \frac{h\nu}{2} + \frac{h\nu}{e^{h\nu/kT} - 1} \tag{7.36}$$

Einstein employed the simple model of all N atoms having the same frequency ν_E. This gives for energy

$$E = 3N\left(\frac{h\nu_E}{2} + \frac{h\nu_E}{e^{h\nu_E/kT} - 1}\right) \tag{7.37}$$

and the specific heat is

$$C_V = \left(\frac{\partial E}{\partial T}\right)_V = 3Nk\left(\frac{h\nu_E}{kT}\right)^2 \frac{e^{h\nu_E/kT}}{(e^{h\nu_E/kT} - 1)^2}$$

$$= 3Nk\left(\frac{\theta_E}{T}\right)^2 \frac{e^{\theta_E/T}}{(e^{\theta_E/T} - 1)^2} \tag{7.38}$$

where

$$\theta_E = \frac{h\nu_E}{k} \tag{7.39}$$

and is called *Einstein characteristic temperature.*

At high temperatures, $T >> \theta_E$, the specific heat $C_V \to 3Nk$, the Dulong and Petit law. At low temperatures, $T << \theta_E$, the specific heat $C_V \to 3Nk(\theta_E/T)^2\, e^{-\theta_E/T}$ and goes to zero at very low temperatures. For many solids, θ_E is about 300 K, and Dulong and Petit law is exhibited at room temperatures, but solids having higher values of θ_E exhibit the law only at high temperatures. The use of quantized vibrations to calculate the thermodynamic quantities for a solid, had a historical importance. However, at low temperatures, the specific heat is proportional to T^3, i.e. decreases more gradually than the exponential decrease indicated by Einstein theory.

In the Debye model, since solids are elastic, elastic waves must be thermally excited. The wave equation is

$$\nabla^2 u = \frac{1}{c^2}\frac{d^2 u}{dt^2} \tag{7.40}$$

Taking the solid to be parallelopiped of size L_x, L_y, L_z (the volume $V = L_x\, L_y\, L_z$), the solution is

$$u = A \sin\frac{n_x \pi x}{L_x} \sin\frac{n_y \pi y}{L_y} \sin\frac{n_z \pi z}{L_z} \cos 2\pi\, \nu t \tag{7.41}$$

The wave vector is related to the overtones by

$$k_\alpha = \frac{n_\alpha \pi}{L_\alpha}, \quad \alpha = x, y, z \tag{7.42}$$

The number of frequencies is very large and hence n_α may be taken as a continuous variable. The number of allowed modes between n_α and $n_\alpha + dn_\alpha$ is

$$dn_x\, dn_y\, dn_z = \frac{L_x L_y L_z}{\pi^3} dk_x\, dk_y\, dk_z = \frac{V}{\pi^3} dk_x\, dk_y\, dk_z \tag{7.43}$$

The frequency of the wave

$$\nu^2 = \frac{k_x^2 + k_y^2 + k_z^2}{4\pi^2} c^2 = \frac{k^2 c^2}{4\pi^2} \tag{7.44}$$

Changing the volume in k-space to spherical polar coordinates,

$$dk_x \, dk_y \, dk_z = 4\pi k^2 dk$$

The number of points having positive values of n_x, n_y, n_z in this volume is $(1/8)(V/\pi^3)$ $4\pi k^2 dk$, since only one of eight octants around a point contains positive values of n_x, n_y, n_z. Since each point determines one mode of vibration, the number of modes using Eq. (7.44) is given by

$$g(\nu)d\nu = \frac{4\pi V}{c^3} \nu^2 \, d\nu \tag{7.45}$$

In an isotropic solid, longitudinal sound waves can be propagated with velocity c_l and also transverse waves with two independent directions of polarization and equal velocities c_t. Hence

$$g(\nu)d\nu = 4\pi V \left(\frac{1}{c_l^3} + \frac{2}{c_t^3} \right) \nu^2 \, d\nu \tag{7.46}$$

In terms of ω, we have

$$g(\omega)d\omega = \frac{V}{2\pi^2} \left(\frac{1}{c_l^3} + \frac{2}{c_t^3} \right) \omega^2 \, d\omega \tag{7.46a}$$

Debye suggested that this distribution of frequencies can be used for crystalline solids as well and the total number of degrees of freedom $3N$ be determined by setting an upper limit of frequency ν_D (called Debye frequency) to the distribution, i.e.

$$\int_0^{\nu_D} g(\nu)d\nu = 4\pi V \left(\frac{1}{c_l^3} + \frac{2}{c_t^3} \right) \int_0^{\nu_D} \nu^2 d\nu = 3N \tag{7.47}$$

This gives the Debye frequency ν_D as

$$\nu_D^3 = \frac{9N}{4\pi V} \left(\frac{1}{c_l^3} + \frac{2}{c_t^3} \right)^{-1} \tag{7.48}$$

Each mode of frequency contributes $(1/2)h\nu + [h\nu/(e^{h\nu/kT} - 1)]$ to the energy, hence

$$E = E_0 + 9NkT \left(\frac{T}{\theta_D} \right)^3 \int_0^{\theta_D/T} \frac{x^3 \, dx}{e^x - 1} \tag{7.49}$$

where $x = h\nu/kT$, $\theta_D = h\nu_D/k$ is called *Debye characteristic temperature* and E_0 is the ground state energy of the solid.

For high temperatures, i.e. $T >> \theta_D$, we have $e^x - 1 \simeq x$ and

$$E - E_0 = 3NkT \tag{7.50}$$

whence

$$C_V = \frac{\partial E}{dT} = 3Nk \tag{7.51}$$

the Dulong and Petit law.

For low temperatures, i.e. $T \ll \theta_D$, the upper limit in the integral in Eq. (7.49) can be replaced by infinity and using Eq. (7.9), we have

$$\int_0^{\infty} \frac{x^3 dx}{e^x - 1} = \Gamma(4)\,\zeta(4) = \frac{\pi^4}{15}$$

so that

$$E - E_0 = \frac{3}{5} Nk\pi^4 \frac{T^4}{\theta_D^3} \tag{7.52}$$

and

$$C_V = \frac{12}{5} Nk\pi^4 \frac{T^3}{\theta_D^3} \tag{7.53}$$

i.e. the specific heat at low temperature varies as T^3.

For temperature variation of specific heat, we first define the Debye function

$$F\left(\frac{\theta_D}{T}\right) = \frac{3}{(\theta_D/T)^3} \int_0^{\theta_D/T} \frac{x^3 dx}{e^x - 1} \tag{7.54}$$

This gives

$$E - E_0 = 3NkT\, F\left(\frac{\theta_D}{T}\right)$$

Differentiating with respect to temperature, we have

$$C_V = 3Nk\, F\left(\frac{\theta_D}{T}\right) - 3Nk\left(\frac{\theta_D}{T}\right) F'\left(\frac{\theta_D}{T}\right) \tag{7.55}$$

At low temperatures, (θ_D/T) is large and the upper limit in the integral of Eq. (7.54) can be replaced by infinity, giving Eq. (7.53). At high temperatures, x is small and $F = 1$ to a first approximation, giving Eq. (7.51). The above gives the temperature dependence of C_V.

The agreement with experimental observations is good. For this reason, the Debye model has found widespread use. But more accurate experiments much later revealed that θ_D is not constant; it varies with temperature. θ_D is fairly constant at the limits of low and high temperatures but has a minimum at around 10 K. Also, the limiting values of θ_D at $T \to 0$ and $T \to \infty$ are not equal. It was also found that θ_D *elastic* and θ_D *calorimetric* agree only at very low temperatures.

Let us examine the normal modes of vibrations of a crystal lattice by taking a simple model, given by Born and Karman, of an infinite, linear, diatomic chain. Such a model includes all the special features of the vibrations of a crystal lattice.

Consider an infinite chain of masses M and m ($M > m$) spaced a apart, masses M occupying odd lattice points and masses m occupying even lattice points. The equations of motion under the assumption of nearest neighbour interactions are

$$m\ddot{u}_{2n} = \beta(u_{2n+1} + u_{2n-1} - 2u_{2n})$$

and

$$M\ddot{u}_{2n+1} = \beta(u_{2n+2} + u_{2n} - 2u_{2n+1}) \tag{7.56}$$

where β is the force constant. We look for the solutions of the form

$$u_{2n} = \xi\, e^{i(\omega t + 2nka)}$$

and

$$u_{2n+1} = \eta\, e^{i[\omega t + (2n+1)ka]} \tag{7.57}$$

which on substitution leads to

$$-\omega^2 m\xi = \beta\eta(e^{ika} + e^{-ika}) - 2\beta\xi$$

and

$$-\omega^2 M\eta = \beta\xi(e^{ika} + e^{-ika}) - 2\beta\eta \tag{7.58}$$

Equation (7.58) has a non-trivial solution only when

$$\begin{vmatrix} 2\beta - m\omega^2 & -2\beta\cos ka \\ -2\beta\cos ka & 2\beta - M\omega^2 \end{vmatrix} = 0 \tag{7.59}$$

i.e.

$$\omega^2 = \beta\left(\frac{1}{m} + \frac{1}{M}\right) \pm \beta\left[\left(\frac{1}{m} + \frac{1}{M}\right)^2 - \frac{4\sin^2 ka}{mM}\right]^{1/2} \tag{7.60}$$

For small values of k, the two roots are

$$\omega^2 = 2\beta\left(\frac{1}{m} + \frac{1}{M}\right) \tag{7.61}$$

and

$$\omega^2 = \frac{2\beta}{M + m} k^2 a^2 \tag{7.62}$$

If the second root given by Eq. (7.62) is used, we find $\xi \simeq \eta$, i.e. the neighbouring atoms move together as in acoustical waves of velocity $[2\beta a^2/(m+M)]^{1/2}$. This branch is called acoustical branch. If the first root given by Eq. (7.61) is used, we get $\xi \simeq -(M/m)\eta$, i.e. the neighbouring atoms move against each other, keeping the centre of mass fixed. If the two atoms have opposite charges, it will be a fluctuating dipole, which has strong interaction with electromagnetic waves. For this reason, it is called optical branch. The frequency in this case is a constant (equal to $[2\beta(M + m)/mM]^{1/2}$) as $k \to 0$, in contrast to the acoustical branch.

The variation of ω with k as given by Eq. (7.60) is shown in Fig. 7.2.

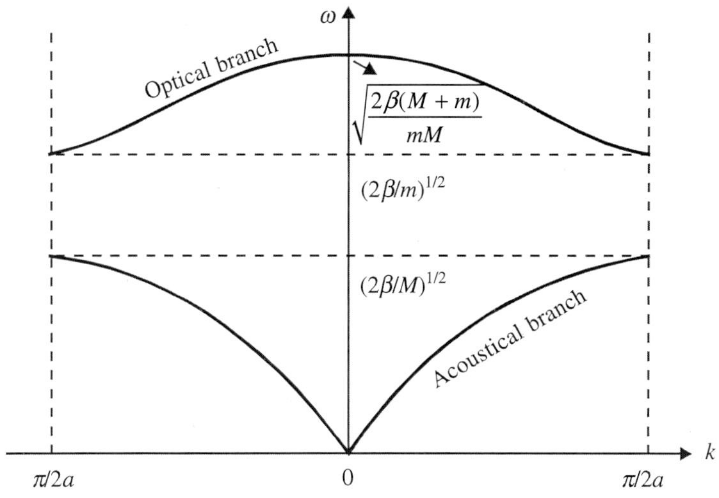

Figure 7.2 Dispersion curve for linear diatomic chain.

In Debye model, ω and k have a linear relation, whereas the diatomic chain has vibrations of both acoustical and optical modes and the phase velocity ω/k in both varies with k, i.e. there is dispersion of velocity. These features are retained in the vibration of three-dimensional crystals as well. The equivalence of a linear harmonic oscillator of mass M and the vibrational mode of a simple one-dimensional lattice of identical atoms of mass m can be easily established. If y is the displacement of the oscillator, its energy is

$$E_{\text{osc}} = \frac{1}{2}M\left(\frac{dy}{dt}\right)^2 + \frac{1}{2}M\omega^2 y^2 \tag{7.63}$$

Let us consider a vibration mode of the lattice corresponding to a standing wave $u_n = \sin nka \cos \omega t$. The kinetic energy is

$$E_{\text{kin}} = \frac{1}{2}m\sum_n\left(\frac{du_n}{dt}\right)^2 = \frac{1}{2}m\,\omega^2 \sin^2 \omega t \sum_n \sin^2 nka \tag{7.64}$$

The potential energy E_{pot} is given by

$$-\frac{\partial E_{\text{pot}}}{\partial u_n} = m\,\frac{d^2u_n}{dt^2} = \beta\sum_n (u_{n+1} + u_{n-1} - 2u_n) \tag{7.65}$$

which yields

$$E_{\text{pot}} = 2\beta \sin^2 ka/2 \cos^2 \omega t \sum_n \sin^2 nka \tag{7.66}$$

The solution corresponding to Eq. (7.60) in this case is

$$\omega^2 = \frac{4\beta}{m} \sin^2 ka/2$$

Substitution of the above gives

$$E_{\text{pot}} = \frac{1}{2} m\omega^2 \cos^2 \omega t \sum_n \sin^2 nka \tag{7.67}$$

From Eqs. (7.64) and (7.67), we get

$$E_{\text{vib}} = \frac{1}{2} m\omega^2 \sum_n \sin^2 nka \tag{7.68}$$

If we use

$$y = \left(\frac{m}{M} \sum_n \sin^2 nka \right)^{1/2} \cos \omega t \tag{7.69}$$

E_{vib} and E_{osc} become identical, i.e. the lattice vibrations can be quantized in a way similar to that of an oscillator. The quantized form of lattice vibrations are called phonons by analogy with photons, the quantized form of electromagnetic waves. The phonons are quasi-particles propagated through the lattice with definite energies and directions of motion. The phonon energy is related to frequency ω by

$$\varepsilon = \hbar\omega \tag{7.70}$$

in the same way as for photons (light quanta). The quasi-momentum $\vec{p}$ is determined by wave vector $\vec{k}$

$$\vec{p} = \hbar\vec{k} \tag{7.71}$$

The velocity of a phonon is given by the group velocity of the corresponding classical waves

$$\vec{c} = \frac{\partial \omega}{\partial \vec{k}} = \frac{\partial \varepsilon(\vec{p})}{\partial \vec{p}} \tag{7.72}$$

which is the usual form of relation between energy, momentum and velocity of particles.

The free wave propagation in harmonic approximation is replaced by free movement of non-interacting phonons in the quantum picture. Various elastic and non-elastic collision processes occur, that establish thermal equilibrium in the phonon gas. The laws of conservation of energy and quasi-momentum must be satisfied in all such processes. Any number of phonons may be created simultaneously in the lattice, i.e., any number of phonons may be in each of the phonon quantum state. Thus, the phonon gas obeys Bose–Einstein statistics. The number of phonons is determined by the equilibrium condition, so its chemical potential is zero (as is the case with photon gas).

For n_i phonons in the ith quantum state, the total energy (leaving the term representing the ground state energy of the solid E_0) is

$$E = \sum_{i=1}^{3N} n_i \hbar \omega_i \tag{7.73}$$

The partition function is given by

$$Q = \sum_{\{n_i\}} e^{-E(n_i)/kT} = \prod_{i=1}^{3N} \frac{1}{1 - e^{-\hbar\omega_i/kT}} \tag{7.74}$$

Hence

$$\ln Q = -\sum_{i=1}^{3N} \ln(1 - e^{-\hbar\omega_i/kT}) \tag{7.75}$$

The average occupation number is obtained as in Eq. (7.4) with $\mu = 0$ as

$$\langle n_i \rangle = \frac{1}{e^{\hbar\omega_i/kT} - 1} \tag{7.76}$$

The total energy is

$$E = \sum_{i=1}^{3N} \langle n_i \rangle \hbar\omega_i = \sum_{i=1}^{3N} \frac{\hbar\omega_i}{e^{\hbar\omega_i/kT} - 1} \tag{7.77}$$

The above can be written (changing from summation to integration over a continuum of phonon states and using the Debye model), as

$$E = \frac{V}{2\pi^2}\left(\frac{1}{c_l^3} + \frac{2}{c_t^3}\right)\int_0^{\omega_D} \frac{\hbar\omega}{e^{\hbar\omega/kT} - 1}\,\omega^2 d\omega = 9NkT\left(\frac{T}{\theta_D}\right)^3 \int_0^{\theta_D/T} \frac{x^3 dx}{e^x - 1} \tag{7.78}$$

where $x = \hbar\omega/kT$ and $\theta_D = \hbar\omega_D/k$, and the integration is on all values of ω in one reciprocal lattice cell.

The free energy is given by

$$A = -kT\ln Q = kT\sum_{i=1}^{3N} \ln(1 - e^{-\hbar\omega_i/kT})$$

$$\equiv kT\sum_{i=1}^{3N} \int \ln(1 - e^{-\hbar\omega/kT})\frac{V}{2\pi^2}\left(\frac{1}{c_l^3} + \frac{2}{c_t^3}\right)\omega^2 d\omega \tag{7.79}$$

The pressure is

$$P = -\left(\frac{\partial A}{\partial V}\right)_T = -\int \frac{h\nu}{e^{h\nu/kT} - 1}\frac{1}{\nu}\frac{\partial \nu}{\partial V}\,g(\nu)\,d\nu = \frac{\gamma E}{V} \tag{7.80}$$

where $\gamma = -[(d \ln \nu)/(d \ln V)]$ and is called *Gruneisen constant.*

It can be seen that

$$\left(\frac{\partial P}{\partial T}\right)_V = \frac{\gamma}{V}\left(\frac{\partial E}{\partial T}\right)_V = -\frac{(\partial V/\partial T)_P}{(\partial V/\partial P)_T} = \frac{\alpha_V}{K_T}$$

where α_V is the coefficient of volume expansion and K_T is the isothermal compressibility. This gives

$$\gamma = \frac{\alpha_V V}{K_T C_V} \tag{7.81}$$

As long as the number of phonons in an assembly is not very large, they do not interact and the assembly may be taken as an ideal phonon gas.

The frequency spectrum of the phonons allows us to calculate the thermodynamic quantities. In a solid angle $d\Omega$, the number of allowed frequencies is

$$\left(\frac{\partial\Omega}{4\pi}\right)4\pi Vp^2 dp = Vk^2\left(\frac{dk}{d\omega}\right)d\omega\, d\Omega$$

Thus, the contribution to $g(\omega)$ is proportional to $k^2/(d\omega/dk)$ in contrast to ω^2/c^3 in the Debye model [see Eq. (7.45)]. In a lattice, $g(\omega)$ starts proportional to ω^2 at $\omega \to 0$ because of low frequency acoustical phonons. As frequency increases, $d\omega/dk$ decreases and the contribution to $g(\omega)$ becomes larger than ω^2, reaches a peak, followed by a minimum where contribution from optical phonons come into picture, giving rise to another peak of much higher value, and then finally falling off. These features were first pointed out by Blackman in 1937. The contribution to specific heat of $g(\nu)$ can now be understood. At low temperatures, $g(\nu)$ varies as ν^2, as in Debye model, and hence $C_V \propto T^3$, the effective Debye temperature is a constant. As temperature increases, $g(\nu)$ becomes larger, so C_V becomes larger, which means that the effective θ_D will decrease, will show a minimum corresponding to the peak in $g(\nu)$ and then increase further. At high temperatures, the entire $g(\nu)$ participates in thermal agitation and the resulting specific heat corresponds to a limiting value of effective θ_D. The lattice theory thus explains the dispersion, the T^3-law at low temperatures, the relationship between θ_D elastic and θ_D calorimetric and the temperature variation of θ_D, all at one stroke.

The specific heat obtained here is the contribution of lattice part only. For metals, the contribution from electronic part (electron gas) has to be added for the total specific heat.

EXERCISES

1. Derive Bose–Einstein distribution law using grand canonical ensemble. Obtain the total number of particles in the gas by normalizing the distribution function.
2. Show that for an ideal monoatomic Bose-Einstein gas, $PV = (2/3)\,E$ and the quantal correction to the pressure in the equation of state is negative.
   Hint: See hints in exercises 1 and 2 of Chapter 6.
3. Discuss Bose–Einstein condensation and find the degeneracy temperature at which the condensation starts.
4. Discuss the specific heat of an ideal Bose gas and explain the kink it shows at the degeneracy temperature.
5. Assume photon gas to be an ideal Bose gas and derive Planck formula and Stefan law.

6. Obtain Wien displacement law from Planck formula.
Hint: The frequency $\nu_m = \nu$, for which the maximum in spectral distribution occurs, is obtained from the condition

$$\frac{d}{d\nu}\left(\frac{\nu^3}{e^{h\nu/kT}-1}\right)_{\nu=\nu_m} = 0 \quad \therefore\ 3\,e^{h\nu_m/kT} - \frac{h\nu_m}{kT}\,e^{h\nu_m/kT} = 3$$

The solution is $h\nu_m/kT \simeq 2.8215$. This means ν_m/T or $\lambda_m T$ is constant, which is Wien displacement law.

7. Calculate the energy per unit volume and pressure of the photon gas by evaluating the partition function.
Hint: For photon gas, $\mu = 0$ and hence there is no difference between Ξ and Q. We have

$$\ln \Xi = \ln Q = -\sum_i \ln(1 - e^{-h\nu_i/kT})$$

The number of states between ν frequencies and $\nu + d\nu$, considering the two polarizations is $(8\pi V/c^3)\nu^2\, d\nu$. Hence the replacement for $\sum_i$ is $\int \frac{8\pi V}{c^3}\nu^2\, d\nu$. Then,

$$\ln \Xi = \ln Q = -\frac{8\pi V}{c^3}\int_0^\infty \nu^2 \ln(1 - e^{-h\nu/kT})\,d\nu$$

Integrating by parts, we get

$$\ln \Xi = \ln Q = -\frac{8\pi V}{c^3}\frac{h}{kT}\int_0^\infty \frac{\nu^2\, d\nu}{e^{h\nu/kT}-1} = \frac{8\pi^5 V}{45}\left(\frac{kT}{hc}\right)^3$$

$$E = kT^2\left(\frac{\partial \ln Q}{\partial T}\right)_V = \frac{8\pi^5 V}{15c^3h^3}k^4T^4 \quad \text{or} \quad u = \frac{E}{V} = \frac{8\pi^5 k^4 T^4}{15c^3h^3}$$

$$P = kT\frac{\ln \Xi}{V} = \frac{8\pi^5 k^4 T^4}{45c^3h^3} = \frac{1}{3}u = \frac{1}{3}\frac{E}{V}$$

Note that $PV = 1/3E$, unlike the relation $PV = 2/3E$ for material particles.

8. Discuss Debye theory of specific heat of solids. What are its shortcomings?
9. What are phonons? How does the study of phonon dispersion curve explain the shortcomings of the Debye model?

8 Imperfect Gases

CLASSICAL CLUSTER EXPANSION

At low densities (limiting case), all gases approach the perfect gas behaviour

$$PV = NkT \text{ or } P = \rho kT \tag{8.1}$$

It is true for a gas in which the interparticle interaction potential could be neglected, owing to the large interparticle separations. This results in almost no effect of one particle on the other.

For a perfect gas, we can write

$$\mathcal{Z}_N = V^N$$

and

$$Q_N = \frac{q^N}{N!}$$

When the density of the gas increases, the particles come closer on the average and the interparticle potential is not negligible. This results in the configuration integral not being V^N and the equation of state Eq. (8.1) does not hold. Experimentally, deviations from perfect gas behaviour are well known. A semi-empirical relation, proposed by van der Waals in 1873,

$$\left(P + \frac{a}{V^2}\right)(V - b) = NkT \tag{8.2}$$

gives good approximation to observations, even down to the volume of the condensed phase.

The most fundamental and theoretically sound virial equation of state, proposed by Thiesen and developed by Kamerlingh Onnes, which expresses the deviations from the ideal gas behaviour is

$$\frac{P}{kT} = \rho + B_2(T)\rho^2 + B_3(T)\rho^3 + \cdots \tag{8.3}$$

The quantities $B_2(T)$, $B_3(T)$, ... are called second, third, ... virial coefficients, which depend only on the temperature and on the particular gas under consideration. We shall derive expressions for the virial coefficients in terms of interparticle potentials.

We start with the partition function of the classical canonical ensemble

$$Q_N = \frac{1}{h^{3N}\,N!}\int\cdots\int e^{-H/kT}\,d\vec{p}_1\ldots d\vec{p}_N\;d\vec{r}_1\ldots d\vec{r}_N \tag{8.4}$$

where

$$H = \sum_{i=1}^{N}\frac{\vec{p}_i^{\,2}}{2m} + U(\vec{r}_1,\ldots,\vec{r}_N) \tag{8.5}$$

Substituting from Eq. (8.5) in Eq. (8.4) and carrying out the momentum integration, we have

$$Q_N = \frac{Z_N}{N!\,\lambda^{3N}} \tag{8.6}$$

where

$$\lambda = \frac{h}{(2\pi mkT)^{1/2}} \tag{8.7}$$

and

$$Z_N = \int\cdots\int e^{-U(\vec{r}_1,\ldots,\vec{r}_N)}\,d\vec{r}_1\ldots d\vec{r}_N \tag{8.8}$$

we assume that the potential energy $U(\vec{r}_1,\ldots,\vec{r}_N)$ can be written as the sum of potentials due to particle pairs, i.e.

$$U(\vec{r}_1,\ldots,\vec{r}_N) = \sum_{1\le i<j\le N} u(r_{ij}) \tag{8.9}$$

where $u(r_{ij})$ is the potential of the interparticle force between particles i and j expressed as a function of the distance r_{ij} between them. $u(r_{ij})$ approaches zero for large values of r_{ij}.

We digress here to obtain in a simple manner the van der Waals equation of state from general statistical considerations. The interparticle forces are strongly repulsive at short distances (this prevents the collapse of the solid into a degenerate mass) and weakly attractive at larger distances (this is responsible for condensation of the gas into liquids and solids). The short-range repulsion is further assumed to be hard-sphere repulsion (potential is infinite in a sphere of a particular radius r_0). Then,

$$\exp\left(\frac{-U}{kT}\right) = \exp\left(\frac{-\sum u(r_{ij})_{\text{attr}}}{kT}\right)\exp\left(\frac{-\sum u(r_{ij})_{\text{rep}}}{kT}\right)$$

which gives

$$\int\cdots\int \exp\left(\frac{-U}{kT}\right)d\vec{r}_1\cdots d\vec{r}_N = \int\cdots\int \prod_{i<j} S(r_{ij})\exp\left(\frac{-\sum u(r_{ij})_{\text{attr}}}{kT}\right)d\vec{r}_1\cdots d\vec{r}_N$$

where $S(r_{ij})$ is a step function.

In the effective range of weak attraction, it depends on the number of particles being attracted and also on the number of particles attracting. Hence we can put on the average,

$$\langle u(r_{ij})_{\text{attr}}\rangle = -CN\frac{N}{V}$$

where C is a constant of proportionality and the –ve sign arises from attraction. The integral $\int\cdots\int \prod_{i<j} S(r_{ij})\,d\vec{r}_1\cdots d\vec{r}_N$ will be V^N if the step function was not present. Its presence will give $(V - b)^N$, b being the excluded volume equal to $(1/2)N[4\pi/3(2r_0)^3]$. This gives

$$Q_N = \frac{1}{N!\,\lambda^{3N}}(V-b)^N e^{CN^2/VkT}$$

Since $A = -kT \ln Q_N$, we have

$$P = -\left(\frac{\partial A}{\partial V}\right)_T = \frac{NkT}{V-b} - \frac{CN^2}{V^2}$$

which gives the van der Waals equation of state

$$\left(P + \frac{CN^2}{V^2}\right)(V - b) = NkT$$

We now return to Eq. (8.9) and introduce Mayer function defined by

$$f_{ij} \equiv f(r_{ij}) = e^{u(r_{ij})/kT} - 1 \tag{8.10}$$

This gives

$$e^{-U/kT} = e^{-\sum u(r_{ij})/kT} = \prod e^{-u(r_{ij})/kT} = \prod_{1\le i<j\le N}(1+f_{ij}) \tag{8.11}$$

The product on expansion gives

$$e^{-U/kT} = 1 + \sum f_{ij} + \sum f_{ij}f_{kl} + \cdots \tag{8.12}$$

The introduction of Mayer function felicitates a more clear exposition of interparticle forces. In the absence of interparticle forces, all f's are zero and we have $Z_N = V^N$. The f's represent the non-zero contribution to $e^{-U/kT}$ of the interparticle forces. The general form of $u(r)$ and $f(r)$ is shown in Fig. 8.1 and Fig. 8.2. It is seen that f_{ij} is zero, except when r_{ij} is small, of the order of several particle diameters.

We thus have

$$Z_N = \int d\vec{r}_1,\cdots d\vec{r}_N \prod_{i<j}(1+f_{ij}) = \int d\vec{r}_1\cdots d\vec{r}_N\left(1 + \sum f_{ij} + \sum f_{ij}f_{kl} + \cdots\right) \tag{8.13}$$

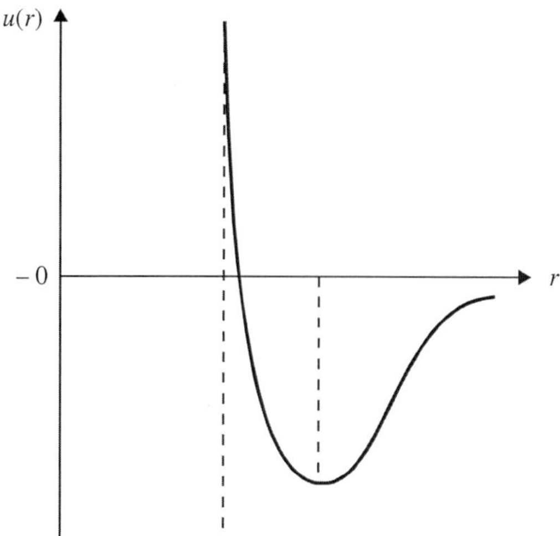

Figure 8.1 Variation of pair potential with interparticle separation.

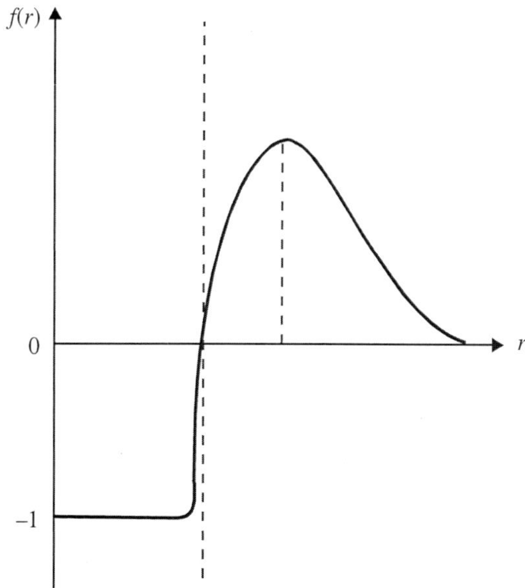

Figure 8.2 Variation of Mayer function with interparticle separation.

A very useful way to account all the terms in the expansion on the right-hand side is to associate a graph with each term.

A N-particle graph consists of N distinct numbered circles with any number of line segments joining a pair of circles.

As an illustration, we give

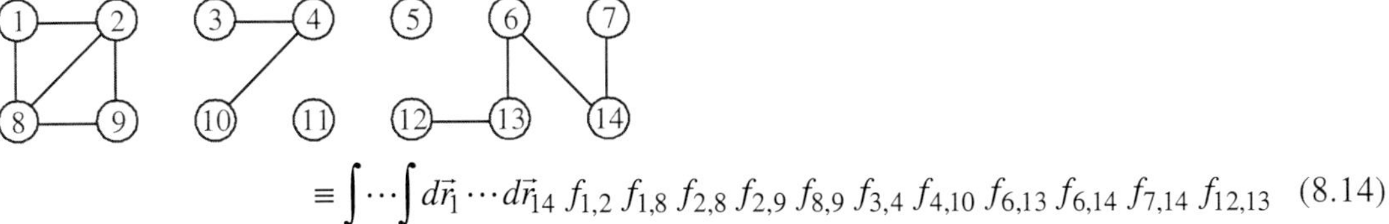

$$\equiv \int \cdots \int d\vec{r}_1 \cdots d\vec{r}_{14}\, f_{1,2}\, f_{1,8}\, f_{2,8}\, f_{2,9}\, f_{8,9}\, f_{3,4}\, f_{4,10}\, f_{6,13}\, f_{6,14}\, f_{7,14}\, f_{12,13} \quad (8.14)$$

With this kind of convention, we have

$$\mathcal{Z}_N = \text{Sum of all distinct } N\text{-particle graphs} \quad (8.15)$$

A j-cluster is defined to be a j-particle graph in which every particle denoted by a circle is joined directly or indirectly to all other $(j-1)$ particles (circles). As an example, we have the following 8-cluster

$$= \int \cdots \int d\vec{r}_1 \cdots d\vec{r}_8\, f_{1,5}\, f_{2,5}\, f_{2,6}\, f_{6,7}\, f_{3,4}\, f_{3,7}\, f_{3,8}\, f_{4,7}\, f_{4,8}\, f_{7,8} \quad (8.16)$$

We now define the cluster integral b_j by

$$b_j = \frac{1}{j!\,V}(\text{sum of all possible } j\text{-clusters}) \quad (8.17)$$

If $S\{m_j\}$ denotes the sum of all graphs corresponding to a set of integers $\{m_j\}$ which specify the collection of graphs, then

$$\mathcal{Z}_N = \sum_{\{m_j\}} S\{m_j\}$$

To find $S\{m_j\}$, let us write down an arbitrary N-particle graph having m_1 unit clusters, m_2 two-particle clusters,..., m_j j-particle clusters and so on. The total number of particles N is then

$$N = \sum_{j=1}^{N} j m_j \quad (8.18)$$

We start by drawing N circles, which can be filled arbitrarily by the numbers 1, 2,..., N. We may permute the numbers to be written in the circles in any way that will lead to a distinct graph. The sum over all permutations gives $S\{m_j\}$, i.e.

$$S\{m_j\} = \sum_p \Big\{\underbrace{[\circ]\cdots[\circ]}_{m_1 \text{ factors}}\Big\}\Big\{\underbrace{[\circ\!-\!\circ]\cdots[\circ\!-\!\circ]}_{m_2 \text{ factors}}\Big\}\Big\{\underbrace{[\triangle]\,[\triangle]\cdots[\triangle]}_{m_3 \text{ factors}}\Big\}\cdots$$

where $\sum\limits_p$ means sum over all permutations:

$$S\{m_j\} = \sum_p \{\circ\}^{m_1}\{\circ\!-\!\circ\}^{m_2}\{\triangle + \triangle + \triangle + \triangle\}^{m_3}\cdots\{\cdots\}^{m_j}\cdots \quad (8.19)$$

Each bracket contains the sum over all j-clusters.

The graphs represent integrals that have values independent of the way in which its circles are numbered. Hence the sum is equal to the number of terms in the sum $\sum_p$ times the value of any term in the sum.

The value of any term in the sum is

$$(1!\,Vb_1)^{m_1}(2!\,Vb_2)^{m_2}\cdots(j!\,Vb_j)^{m_j}\cdots=\prod_{j=1}^{N}(j!\,Vb_j)^{m_j} \tag{8.20}$$

To find the number of terms in the sum $\sum_p$, we note that a permutation of m_j j-clusters amongst themselves does not give a new graph. Hence the $N!$ permutations of N distinguishable particles must be divided by $\Pi m_j!$. It should be noted that indistinguishability has already been taken care of by the factor $N!$ in Eq. (8.6). Further, the permutation of j-particles in j-cluster does not give a new graph. Hence, a further division by $\prod_j (j!)^{m_j}$ is required. This gives the number of terms in the sum as

$$\prod_j \frac{N!}{(j!)^{m_j}\, m_j!} \tag{8.21}$$

Therefore, we have

$$S\{m_j\}=\prod_j (j!Vb_j)^{m_j}\prod_j \frac{N!}{(j!)^{m_j}\, m_j!}=N!\prod_{j=1}^{N}\frac{(Vb_j)^{m_j}}{m_j!} \tag{8.22}$$

Consequently, we have

$$\mathcal{Z}_N=N!\sum_{\{m_j\}}\prod_{j=1}^{N}\frac{(Vb_j)^{m_j}}{m_j!} \tag{8.23}$$

$$Q_N=\frac{1}{\lambda^{3N}}\sum_{\{m_j\}}\prod_{j=1}^{N}\frac{(Vb_j)^{m_j}}{m_j!} \tag{8.24}$$

The theory presented here is applicable to classical monoatomic gas with interparticle interactions expressible in terms of pair potentials. However, developments due to Yang and Lee (*Phys. Rev.* **87**, 404, 1952) extends it to include the liquid phase and a simple description of the condensation phenomenon.

THE VIRIAL EQUATION OF STATE

We start with the grand partition function

$$\Xi=\sum_{N\geq 0}\frac{z^N\mathcal{Z}_N}{N!}$$

and substitute for $\mathcal{Z}_N$ from Eq. (8.23), and use

$$z^N=z^{\sum_j jm_j}=\prod_j (z^j)^{m_j}$$

to obtain

$$\Xi = \prod_{j\geq 1}\left[\sum_{\{m_j\}} \frac{(Vb_j z^j)^{m_j}}{m_j!}\right] = \prod_{j\geq 1} e^{Vb_j z^j} \tag{8.25}$$

This gives

$$\frac{1}{V}\ln \Xi = \sum_j b_j z^j \tag{8.26}$$

Since

$$\frac{P}{kT} = \underset{V\to 0}{\text{Lt}}\ \frac{1}{V}\ln \Xi$$

we get

$$\frac{P}{kT} = \sum_{j\geq 1} b_j z^j \tag{8.27}$$

This gives the expansion of pressure in a power series of fugacity. The series on the right hand side converges rapidly and only a few terms are appreciable.

We are more interested in an expansion of pressure in a power series of number density ρ $(= N/V)$. To do so, we start with

$$\langle N\rangle = N = z\left(\frac{\partial \ln \Xi}{\partial z}\right)_{V,T}$$

Substitution from Eq. (8.26) gives

$$\rho = \frac{N}{V} = \sum_{j\geq 1} jb_j z^j \tag{8.28}$$

This gives the expansion of number density in a power series of fugacity.

To obtain an expansion of pressure in terms of number density, we eliminate fugacity from the Eqs. (8.27) and (8.28). A general method has been given by Kahn.

However, we can find the few initial terms algebraically. To do so, we substitute

$$z = a_1\rho + a_2\rho^2 + a_3\rho^3 + \ldots$$

in Eq. (8.28)

$$\sum_{j\geq 1} jb_j z^j - \rho = 0$$

and equate the coefficients of each power of ρ equal to zero. This will give the a's in terms of b's

$$\begin{aligned}
a_1 &= b_1 = 1\\
a_2 &= -2b_2\\
a_3 &= -3b^3 + 8b_2^2\\
a_4 &= -4b_4 + 30b_2b_3 - 40b_2^2\\
&\ldots \qquad \ldots \qquad \ldots
\end{aligned} \tag{8.29}$$

We then substitute

$$z = \rho - 2b_2\rho^2 + (8b_2^2 - 3b_3)\,\rho^3 + \cdots \tag{8.30}$$

for z in Eq. (8.27) to obtain

$$\frac{PV}{NkT} = 1 - b_2\rho + (4b_2^2 - 2b_3)\,\rho^2 + \cdots \tag{8.31}$$

which is the virial equation of state, which shows the relationship of the virial coefficients and the cluster integrals. Evaluation of the virial coefficients involves only a computation of a number of integrals.

We have useful relations to manipulate, like

$$\Xi = \sum_{N \geq O} Q_N z'^N = \sum_{N \geq O} \frac{Z_N z^N}{N!}$$

which gives

$$z = \frac{Q_1}{Z_1} z' = \frac{Q_1}{V} z' \tag{8.32}$$

Also,

$$\Xi = 1 + \sum_{N \geq 1} \frac{Z_N z^N}{N!} = e^{PV/kT} \tag{8.33}$$

We expand the exponential, substitute for P from Eq. (8.27) and compare the coefficients of like powers of z on two sides, to get

$$\begin{aligned}
b_1 &= (1!V)^{-1} Z_1 = 1 \\
b_2 &= (2!V)^{-1} (Z_2 - Z_1^2) \\
b_3 &= (3!V)^{-1} (Z_3 - 3Z_2Z_1 + 2Z_1^3) \\
b_4 &= (4!V)^{-1} (Z_4 - 4Z_3Z_1 - 3Z_2^2 + 12Z_2Z_1^2 - 6Z_1^4) \\
\cdots\cdots & \quad \cdots\cdots \quad \cdots\cdots \quad \cdots
\end{aligned} \tag{8.34}$$

IRREDUCIBLE CLUSTER INTEGRALS

The cluster integrals b_j defined by Eq. (8.17) contain the sum of all products of f_{ij}'s corresponding to different singly connected cluster diagrams with j-particles. If we examine this closely, we find that the integral can be decomposed into what we call irreducible cluster integrals, β_k. The numerical calculation of β_k is unfortunately simple only for few β_k's, but they have a special significance, because they turn out to be essentially the coefficients in the virial expansion of pressure. An irreducible cluster is defined as any product of f_{ij}'s associated with a doubly connected diagram, except that a cluster of two particles is also irreducible. We illustrate it by the following diagrams.

Irreducible clusters for j = 2, 3, 4 are shown below

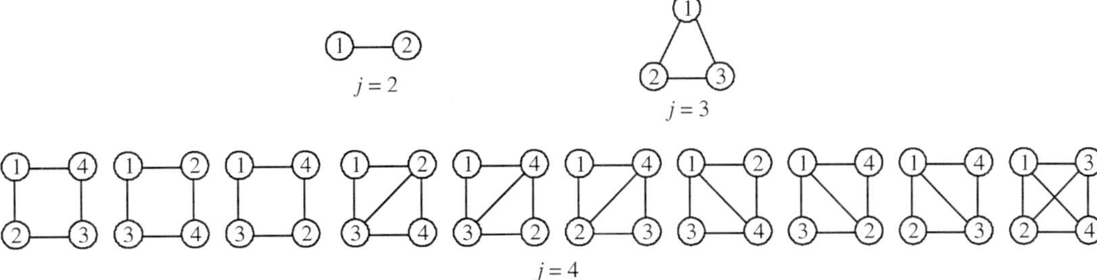

A cluster of 8 molecules shown below has two irreducible clusters with j = 2, one irreducible cluster with j = 3 and one irreducible cluster with j = 4

$$\equiv (f_{2,6})(f_{6,7})(f_{1,2}\ f_{1,5}\ f_{2,5})(f_{3,4}\ f_{3,7}\ f_{3,8}\ f_{4,8}\ f_{7,8})$$

The irreducible cluster integral β_k is defined by

$$\beta_k = \frac{1}{k!V} \int \cdots \int s'_{1,2,\ldots,k+1}\ d\vec{r}_1 \cdots d\vec{r}_{k+1} \tag{8.35}$$

where $s'_{1,2,\ldots,k+1}$ is the sum of all different terms (product of f_{ij}'s) that connect the particles 1, 2,..., k + 1 into an irreducible cluster.

For example $s'_{1,2,3,4}$ is a sum of 10 terms corresponding to the irreducible cluster (for j = 4) in the above graph. Thus,

$$s'_{1,2,3,4} = 3(f_{1,2}\ f_{2,3}\ f_{3,4}\ f_{1,4}) + 6(f_{1,2}\ f_{2,3}\ f_{3,4}\ f_{2,4}\ f_{1,3}) + (f_{1,2}\ f_{2,3}\ f_{3,4}\ f_{1,4}\ f_{1,3}\ f_{2,4})$$

k is called the index of irreducible cluster.

The irreducible cluster integrals β_k are related to the cluster integrals b_j (Kahn, B. Ph.D. Dissertation, Utrecht. 1938, reprinted in *Studies in Statistical Mechanics* edited by Boer, J. de and Uhlenbeck, G.E., North Holland Publishing Co. 1965, Vol. III) by the relation

$$\beta_k = \sum_m (-1)^{\sum_j m_j - 1} \frac{\left(k - 1 + \sum_j m_j\right)!}{k!} \prod_j \frac{(jb_j)^{m_j}}{m_j!} \tag{8.36}$$

where the first summation is over all sets of positive integers or zero, $m = m_2, m_3, \ldots,$ such that

$$\sum_j (j-1) m_j = k \tag{8.37}$$

This gives β_k in terms of b_j.

The relation between the first few b_j's and β_k's can be written with the aid of Eq. (8.36) as

$$
\begin{aligned}
b_2 &= \frac{1}{2}\beta_1 \quad \text{or} \quad \beta_1 = 2b_2 \\
b_3 &= \frac{\beta_1^2}{2} + \frac{\beta_2}{3} \quad \text{or} \quad \beta_2 = 3b_3 - 6b_2^2 \\
b_4 &= \frac{2\beta_1^3}{3} + \beta_1\beta_2 + \frac{1}{4}\beta_3 \quad \text{or} \quad \beta_3 = 4b_4 - 24b_2b_3 + \frac{80}{3}b_2^3 \\
&\cdots \qquad \cdots \qquad \cdots
\end{aligned}
\tag{8.38}
$$

To relate the irreducible cluster integrals to the virial coefficients, we rewrite Eq. (8.31) as

$$\frac{PV}{NkT} = 1 - b_2\rho + (4b_2^2 - 2b_3)\rho^2 + \cdots$$

The above can be expressed in terms β_k's by using Eq. (8.38)

$$\frac{PV}{NkT} = 1 - \frac{1}{2}\beta_1\rho - \frac{2}{3}\beta_2\rho^2 - \cdots \tag{8.39}$$

The complete expression is

$$\frac{PV}{NkT} = 1 - \sum_{k\geq 1} \frac{k}{k+1}\beta_k\rho^k \tag{8.40}$$

If we compare the above with the virial equation of state Eq. (8.3), we get the relation between the virial coefficients and the irreducible cluster integrals

$$B_k = -\frac{k-1}{k}\beta_{k-1} \quad \text{for } k \geq 2 \tag{8.41}$$

Kahn made an alternative derivation relating the irreducible cluster integrals to the virial coefficients by first obtaining the inverses of Eq. (8.28) as

$$z = \rho e^{-\sum_{k\geq 1} \beta_k \rho^k} \tag{8.42}$$

Then, we can write Eq. (8.27)

$$
\begin{aligned}
\frac{P}{kT} &= \int_0^z \left(\sum_{j\geq 1} j b_j z^{j-1} \right) dz = \int_0^z \frac{\rho}{z}\, dz = \int_0^\rho e^{\sum \beta_k \rho^k}\, d\left[\rho e^{-\sum \beta_k \rho^k}\right] \\
&= \int_0^\rho \left(1 - \rho \sum_{j\geq 1} k\beta_k \rho^{k-1} \right) d\rho = \rho - \sum_{k\geq 1} \frac{k}{k+1}\beta_k \rho^k
\end{aligned}
\tag{8.43}
$$

Comparison with virial equation of state Eq. (8.3) gives

$$B_k = -\frac{k-1}{k}\beta_{k-1} \quad \text{for } k \geq 2 \tag{8.44}$$

THE VIRIAL COEFFICIENTS IN CLASSICAL LIMITS

We compare the virial equation of state Eq. (8.3) with Eq. (8.31), and then we have the virial coefficients

$$
\begin{aligned}
B_2(T) &= -b_2 = -(2!V)^{-1}(\mathcal{Z}_2 - \mathcal{Z}_1^2) \\
B_3(T) &= -4b_2^2 - 2b_3 = -\frac{1}{3V^2}\left[V(\mathcal{Z}_3 - 3\mathcal{Z}_2\mathcal{Z}_1 + 2\mathcal{Z}_1^3) - 3(\mathcal{Z}_2 - \mathcal{Z}_1^2)^2\right] \\
&\cdots \qquad \cdots \qquad \cdots
\end{aligned}
\tag{8.45}
$$

We can write

$$
\begin{aligned}
\mathcal{Z}_1 &= \iint d\vec{r} = V \\
\mathcal{Z}_2 &= \iint d\vec{r}_1\, d\vec{r}_2\, e^{-U_2/kT} = \iint d\vec{r}_1\, d\vec{r}_2\, e^{-u(r_{12})/kT} \\
\mathcal{Z}_3 &= \iint d\vec{r}_1\, d\vec{r}_2\, d\vec{r}_3\, e^{-U_3/kT} = \iiint d\vec{r}_1\, d\vec{r}_2\, d\vec{r}_3\, e^{-[u(r_{12})+u(r_{13})u(r_{23})]/kT}
\end{aligned}
\tag{8.46}
$$

We substitute from Eq. (8.46) in Eq. (8.45) to find

$$
B_2(T) = -\frac{1}{2V}\iint (e^{-u(r_{12})/kT} - 1)\, d\vec{r}_1\, d\vec{r}_2 = -\frac{1}{2}\int (e^{-u(r)/kT} - 1)\, d\vec{r} \tag{8.47}
$$

and after a lengthy calculation, we get

$$
B_3(T) = -\frac{1}{3V}\iiint f_{12}\, f_{13}\, f_{23}\, d\vec{r}_1\, d\vec{r}_2\, d\vec{r}_3 \tag{8.48}
$$

We can calculate the virial coefficients if the interparticle potential $u(r)$ is known. The second virial coefficient can be measured easily over a large temperature range and gives useful information about the interparticle interaction. We evaluate the second virial coefficient for few simple examples: (a) a hard-sphere potential, (b) a hard-sphere potential followed by a square well potential and (c) Lennard–Jones potential.

(a) Hard-sphere potential: A hard-sphere potential is defined by

$$
\begin{aligned}
u(r) &= \infty \quad \text{for } 0 < r < a \\
u(r) &= 0 \quad \text{for } r > a
\end{aligned}
$$

The potential has a steep repulsive part in the region $0 < r < a$ and no attractive part (Fig. 8.3). We substitute this potential in Eq. (8.47) for the second virial coefficient, to get

$$
B_2(T) = -\frac{1}{2}\int_0^a (-1)\, 4\pi r^2 dr = \frac{2\pi a^3}{3} \tag{8.49}
$$

This is four times the volume of a sphere of radius $a/2$ and is temperature independent.

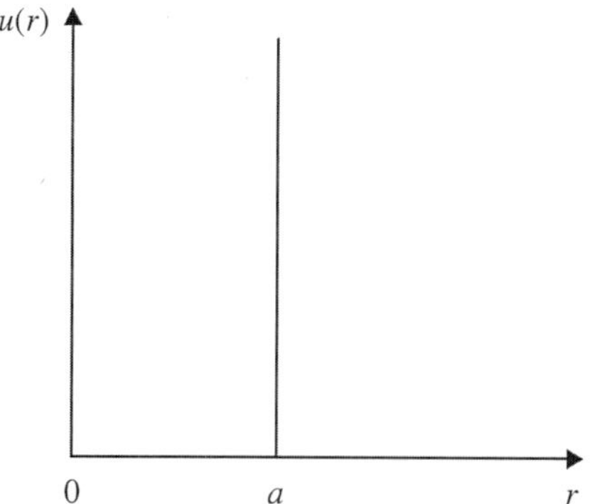

Figure 8.3 Hard-sphere potential.

(b) Hard-sphere with square well potential tail: The potential is defined by

$$u(r) = \infty \quad \text{for } 0 < r < a$$
$$u(r) = -\varepsilon \quad \text{for } a < r < b$$
$$u(r) = 0 \quad \text{for } r > b$$

This potential has a steep repulsive core in the region $0 < r < a$, followed by an attractive well of constant depth in the region $a < r < b$ and is zero beyond b (Fig. 8.4). We substitute this potential in Eq. (8.47) for the second virial coefficient to get

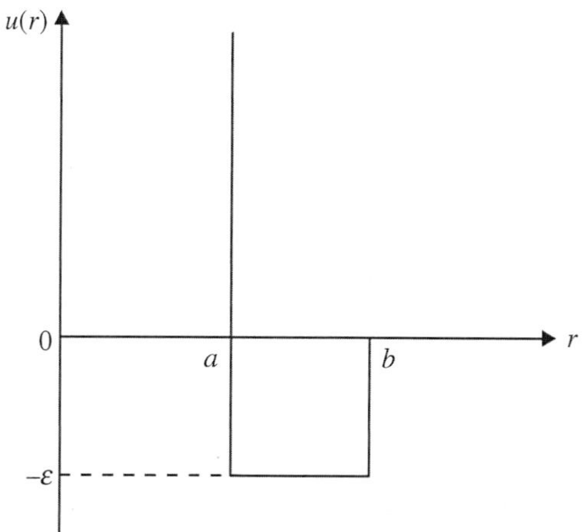

Figure 8.4 Hard-sphere with square well potential tail.

$$B_2(T) = -\frac{1}{2}\int_0^a (-1)4\pi r^2 dr - \frac{1}{2}\int_0^b (e^{\varepsilon/kT} - 1)4\pi r^2 dr$$

$$= \frac{2\pi a^3}{3} - \frac{2}{3}\pi(b^3 - a^3)(e^{\varepsilon/kT} - 1) \tag{8.50}$$

The above reduces to hard-sphere result if $b \to a$ or $\varepsilon \to 0$.

(c) Lennard–Jones potential: The most widely used inter-atomic potential is the Lennard–Jones potential (Fig. 8.5), defined by

$$u(r) = 4\varepsilon\left[-\left(\frac{\sigma}{r}\right)^6 + \left(\frac{\sigma}{r}\right)^{12}\right] \tag{8.51}$$

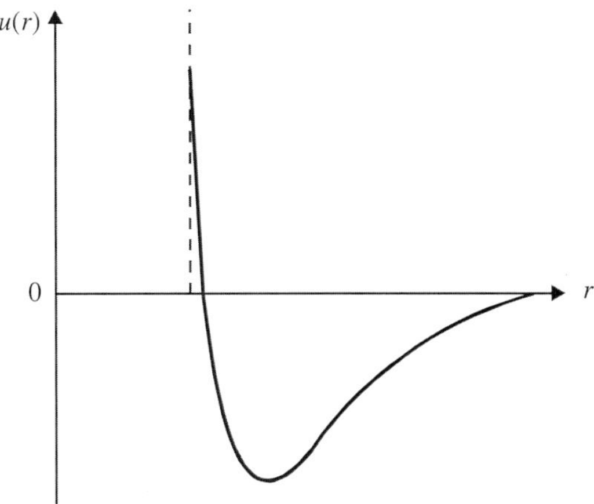

Figure 8.5 Lennard-Jones potential.

with ε = 10.22 K and σ = 2.556 Å. Minimum in the potential occurs at $r = 2^{1/6}\ \sigma$.

The second virial coefficient for this potential is

$$B_2(T) = -\frac{1}{2}\int_0^\infty \left[\exp\left\{-\frac{4\varepsilon}{kT}\left[\left(\frac{\sigma}{r}\right)^{12} - \left(\frac{\sigma}{r}\right)^6\right]\right\} - 1\right] 4\pi r^2 dr$$

The result is

$$B_2(T) = \frac{2}{3}\pi\sigma^3\left(\frac{kT}{\varepsilon}\right)^{-1/4}\left[1.733 - 2.56\left(\frac{kT}{\varepsilon}\right)^{-1/2} - 8.66\left(\frac{kT}{\varepsilon}\right)^{-1} - 4.27\left(\frac{kT}{\varepsilon}\right)^{-3/2} + \cdots\right] \tag{8.52}$$

QUANTUM CLUSTER EXPANSION

Consider a system of N identical particles in a volume V. The quantum partition function is

$$Q_N = T_r e^{-\beta\hat{H}} = \int\cdots\int d\vec{r}_1 \cdots d\vec{r}_N \sum_\alpha \Psi_\alpha^*(\vec{r}_1,\ldots,\vec{r}_N)\, e^{-\beta\hat{H}}\, \Psi_\alpha(\vec{r}_1,\ldots,\vec{r}_N) \tag{8.53}$$

where Ψ_α is a complete set of orthonormal wave functions, symmetrized or antisymmetrized, as appropriate to the system. We define

$$W_N(1,\ldots,N) = N!\lambda^{3N}\sum_\alpha \Psi_\alpha(1,\ldots,N)\, e^{-\beta\hat{H}}\, \Psi_\alpha(1,\ldots,N) \tag{8.54}$$

where $W_N(1,\ldots, N)$ is the probability density matrix and set of coordinates $\{\vec{r}_1,\ldots,\vec{r}_N\}$ has been abbreviated as $\{1,\ldots, N\}$. This gives

$$Q_N = \frac{1}{N!\lambda^{3N}} \int d\vec{r}_1 \ldots d\vec{r}_N W_N(1,\ldots, N) \tag{8.55}$$

The integral on the right-hand side approaches the classical configurational partition function in the limit of high temperatures. We note some properties of $W_N(1,\ldots, N)$.

(i) $W_1(1) = 1.$ (8.56)

(ii) $W_N(1,\ldots, N)$ is a symmetric function of its arguments.

(iii) $W_N(1,\ldots, N)$ is invariant under a unitary transformation of the complete set of wave functions $\{\Psi_\alpha\}$.

(iv) If the coordinates $\vec{r}_1,\ldots,\vec{r}_N$ can be divided in two groups containing A and B coordinates in the group, such that $\vec{r}_i$ and $\vec{r}_j$ belonging to the two different groups satisfy the condition $|\vec{r}_i - \vec{r}_j| \gg \lambda$, then

$$W_N(\vec{r}_1,\ldots,\vec{r}_N) = W_A(\vec{r}_A)\, W_B(\vec{r}_B) \tag{8.57}$$

where $\vec{r}_A$ and $\vec{r}_B$ denote all the coordinates in their groups.

We now proceed with the development due to Kahn and Uhlenbeck (*Physica*, **5**, 399, 1938). If follows from Eq. (8.57) for $|\vec{r}_1 - \vec{r}_2| \to \infty$

$$W_2(1, 2) \to W_1(1)\; W_1(2)$$

We define Ursell function $U_2(1, 2)$ by

$$W_2(1, 2) = W_1(1)\, W_2(2) + U_2(1, 2)$$

and expect that

$$U_2(1, 2) \to 0 \quad \text{as} \quad |\vec{r}_1 - \vec{r}_2| \to \infty$$

Thus, the integral of $U_2(1, 2)$ over $\vec{r}_1$ and $\vec{r}_2$ is the quantum mechanical analogue of the classical 2-cluster.

Let the cluster function $U_j(1,\ldots, j)$ defined by the following scheme.

$$W_1(1) = U_1(1) \doteq 1$$

$$W_2(1, 2) = U_1(1)\; U_1(2) + U_2(1, 2)$$

$$\begin{aligned} W_3(1, 2, 3) = {} & U_1(1)\; U_1(2)\; U_1(3) + U_1(1)\; U_2(2, 3) + U_1(2)\; U_2(3, 1) \\ & + U_1(3)\; U_2(1, 2) + U_3\;(1, 2, 3) \end{aligned} \tag{8.58}$$

$$\ldots \qquad\qquad \ldots \qquad \ldots$$

$$W_N(1, 2,\ldots, N) = \sum_{mj}\sum_{j} \underbrace{[U_1(r)\cdots U_1(r)]}_{m_1 \text{ factors}}\; \underbrace{[U_2(,)\cdots U_2(,)]}_{m_2 \text{ factors}} \;\ldots\;\ldots\; \underbrace{[U_N(,\ldots,)]}_{m_N \text{ factors}}$$

where m_j is zero or a positive integer

and

$$\sum_{j=1}^{N} jm_j = N \tag{8.59}$$

The arguments of U_j have been left blank, which are to be filled by the coordinates $\vec{r}_1,\ldots,\vec{r}_N$ in any order. The sum is over all the distinct ways of filling these blanks.

We solve Eq. (8.58) for $U_j(1,\ldots,j)$ to obtain

$$\begin{aligned}
U_1(1) &= W_1(1) = 1 \\
U_2(1,2) &= W_2(1,2) - W_1(1)\, W_1(2) \\
U_3(1,2,3) &= W_3(1,2,3) - W_2(1,2)\, W_1(3) - W_2(2,3)\, W_1(1) \\
&\quad - W_2(3,1)\, W_1(2) + 2W_1(1)\, W_1(2)\, W_1(3) \\
\cdots & \qquad \cdots \qquad \cdots
\end{aligned} \tag{8.60}$$

We note that $U_j(1,\ldots,j)$ is a symmetric function of its arguments and is determined by all the $W_{N'}$ with $N' \le j$.

The cluster integral b_j is defined by

$$b_j = \frac{1}{j!V}\int\cdots\int d\vec{r}_1 \ldots d\vec{r}_j\, U_j(1,\ldots,j) \tag{8.61}$$

According to Eqs. (8.55) and (8.58), we find that the partition function can be expressed in terms of the cluster integrals. We need to integrate W_N over all coordinates. The result of the integration is the number of terms in the integration times the integral of any term. The number of terms is found in a manner exactly similar to the classical case in Eq. (8.22). Hence

$$\begin{aligned}
\int\cdots\int d\vec{r}_1 \cdots d\vec{r}_n W_N(1,\ldots,N) &= \sum_{\{m_j\}} \frac{N!}{\prod_j (j!)^{m_j} m_j!} \int\cdots\int d\vec{r}_1 \ldots d\vec{r}_N\, [(U_1 \ldots U_1)(U_2 \ldots U_2)\cdots] \\
&= N! \sum_{\{m_j\}} \frac{1}{m_1!}\left[\frac{1}{1!}\int d\vec{r}_1 U_1(1)\right]^{m_1} \frac{1}{m_2!}\left[\frac{1}{2!}\int d\vec{r}_1\, d\vec{r}_2\, U_2(1,2)\right]^{m_2} \cdots \\
&= N! \sum_{\{m_j\}} \prod_j \frac{(Vb_j)^{m_j}}{m_j!}
\end{aligned} \tag{8.62}$$

Substitution in Eq. (8.55) gives

$$Q_N = \frac{1}{\lambda^{3N}} \sum_{\{m_j\}} \prod_{j=1}^{N} \frac{(Vb_j)^{m_j}}{m_j!} \tag{8.63}$$

This is exactly of the same form as that in the classical case Eq. (8.24). Hence the discussion following Eq. (8.24) applies to the quantal case also. In the quantal case, however, the calculation of b_j needs the knowledge of U_j and consequently $W_{N'}$ for

$N' \leq j$. This is in contrast to the classical case in which b_j involves only a number of integrals that can be evaluated. For $j > 1$ in the quantal case, we need to solve a j-body problem which has no finite prescription for $j > 2$.

EXERCISES

1. What are cluster integrals? Express the classical canonical partition function in terms of cluster integrals.
2. Derive the cluster expansion of pressure of a classical gas and hence relate the virial coefficients with the cluster integrals.
3. What are irreducible cluster integrals? How are they related to the virial coefficients?
4. Evaluate the second virial coefficient for a classical hard-sphere gas and also the hand-sphere gas followed by an attractive square well potential.
5. Derive quantal partition function for a canonical ensemble in terms of the cluster integrals.
6. Find $(C_P - C_V)$ for a van der Waals gas.

Hint: $$C_P - C_V = -T\left[\left(\frac{\partial P}{\partial T}\right)_V^2\right]\left(\frac{\partial P}{\partial V}\right)_T = -T\left[\left(\frac{\partial V}{\partial T}\right)_P\right]^2\left(\frac{\partial V}{\partial P}\right)_T$$

van der Waals equation of state is

$$P + \frac{a}{V^2} = \frac{RT}{V - b}$$

$$\therefore \quad \left(\frac{\partial V}{\partial T}\right)_P = \frac{R}{(V-b)\left[\dfrac{RT}{(V-b)^2} - \dfrac{2a}{V^3}\right]}, \quad \left(\frac{\partial P}{\partial V}\right)_T = -\frac{RT}{(V-b)^2} + \frac{2a}{V^3}$$

Substitution gives

$$C_p - C_v = \frac{R}{1 - \dfrac{2a(V-b)^2}{RTV^3}} \simeq R\left(1 + \frac{2a}{RTV}\right)$$

9 Phase Transition and Critical Phenomena

PHASE TRANSITION

Matter exists in three phases—solid, liquid and gas. States of matter that exist simultaneously in equilibrium with one another and in contact, are called different phases. Transition from one phase to another phase is called *phase transition*. The transition takes place at constant temperature and pressure. This pressure $P(T)$ at the temperature T is called vapour pressure. To understand the transition let us refer to the P–T and P–V diagrams (Fig. 9.1 and Fig. 9.2) of a typical system.

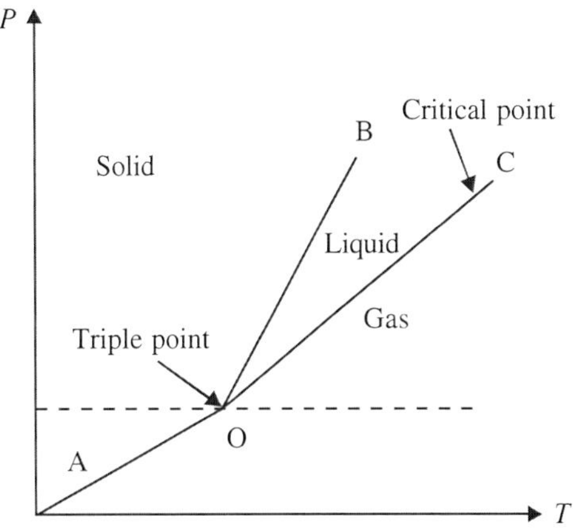

Figure 9.1 P–T diagram.

If we write $g_2 - g_1 = \Delta g$, then

$$\frac{(\partial \Delta g/\partial T)_P}{(\partial \Delta g/\partial P)_T} = -\frac{s_2 - s_1}{v_2 - v_1} = -\left(\frac{\partial P}{\partial T}\right)_{\Delta g} \tag{9.7}$$

Since entropy is always positive, g is a decreasing function of T.

This gives

$$\left(\frac{\partial P}{\partial T}\right)_{\Delta g} = \frac{s_2 - s_1}{v_2 - v_1} \tag{9.8}$$

Since the derivative is defined as

$$\frac{dP}{dT} = \left(\frac{\partial P}{\partial T}\right)_{\Delta g = 0} \tag{9.9}$$

we get

$$\frac{dP}{dT} = \frac{1}{T}\frac{T(s_2 - s_1)}{v_2 - v_1} = \frac{L}{T(v_2 - v_1)} \tag{9.10}$$

where $L = T(s_2 - s_1)$ is called the latent heat of transition. This is known as *Clausius–Clapeyron equation* that gives the change of vapour pressure with temperature.

In addition to this, there is a phase transition of second order in which the first derivatives of Gibbs potential at transition point is equal, i.e. $s_2 = s_1$ and $v_2 = v_1$. Ehrenfest defines a phase transition to be of nth order if at the transition point

$$\frac{\partial^n g_1}{\partial T^n} \neq \frac{\partial^n g_2}{\partial T^n} \quad \text{and} \quad \frac{\partial^n g_1}{\partial P^n} \neq \frac{\partial^n g_2}{\partial P^n} \tag{9.11}$$

whereas all lower derivatives are equal.

No discontinuous change of state at the phase transition of second kind implies that the thermodynamic functions like energy, entropy and volume vary continuously as the transition point is passed. Hence there is no absorption or evolution of heat at the transition. But the derivatives of the thermodynamic quantities like the specific heat, the thermal expansion coefficient, the compressibility, etc. are discontinuous at the transition point.

The transition between phases of different symmetry (different crystal modifications) does not occur in a continuous manner. The body has one symmetry or the other in every state and hence can be assigned to one phase or the other. The transition between different crystal modifications are brought about by phase transition in which there is an abrupt rearrangement of the crystal lattice. We take the example of $BaTiO_3$, which has cubic lattice at high temperatures. As the temperature is lowered below a certain value, the titanium and oxygen atoms begin to move relative to the barium atoms affecting the symmetry of the lattice and it becomes tetragonal in place of cubic. The change in symmetry also may occur due to a change in the ordering of the crystal. Increase of symmetry means disordered (less ordered) state. For example, a completely ordered alloy of CuZn (brass) has a simple cubic Bravis lattice with zinc atoms at the vertices and copper atoms at the centres of the cubic cells, the probabilities of these sites for occupancy by either type of atom is not equivalent. When the alloy becomes disordered, the probabilities of finding an atom of one kind or the other becomes equal at all the sites and the symmetry is raised. Both of these are phase transitions of the second kind. This kind of transition may bring about the change in some other property of symmetry, as is the case in the ferromagnetic substances at Curie point. In this case, the symmetry of the arrangement of elementary magnetic moments in the body is changed. The other examples of the second order phase transition are of liquid helium to superfluid state and that of metal to superconducting state. In both these cases, the body acquires a new property at the transition point.

We have seen that in the second order phase transition, the symmetry of one phase is higher than the other. The more symmetrical phase corresponds to higher temperature. We define an order parameter η such that it takes non-zero values in the phase with less symmetry (ordered phase) and zero in the phase with more symmetry (disordered phase). The phase transition of the second kind has a continuous change of η to zero in contrast to discontinuous change of η in the phase transition of the first kind.

LANDAU THEORY OF SECOND ORDER PHASE TRANSITION

Let the thermodynamic potential G be a function of P, T and η. The variables P and T can be specified arbitrarily but the variable η must be determined from the condition that for a given P and T, G must be minimum in thermal equilibrium. Since the change of state is continuous in second order phase transition, η takes arbitrarily small values near the transition point. In the neighbourhood of the transition point, we expand G (P, T, η) in powers of η

$$G(P,T,\eta) = G_0 + \alpha\eta + A\eta^2 + D\eta^3 + B\eta^4 + \cdots \tag{9.12}$$

the coefficients α, A, D, B, ... are functions of P and T.

The expansion of the type given by Eq. (9.12) does not take into account that the transition point is a singularity of free energy. However, Landau theory assumes the validity of such an expansion under certain conditions. Since the states of $\eta = 0$ and $\eta \neq 0$ are of different symmetry, the first order term in the expansion is zero, i.e., $\alpha = 0$. At the transition point, in the phase with more symmetry, $\eta = 0$ should correspond to a minimum of G. Hence $A > 0$. Looking from the other side of the transition, i.e., the phase with less symmetry, the non-zero values of η correspond to the stable state, i.e., G is minimum. This is possible only if $A < 0$. Hence A is positive on one side of the transition point and negative on the other side; it must vanish at the transition point itself. A plot of $G(\eta)$ is shown for $A > 0$ and $A < 0$ in Fig. 9.5.

At the transition point, which is a stable state, G as a function of η, is a minimum at $\eta = 0$. Hence necessarily we have at transition point

$$A_c = 0,\ D_c = 0,\ B_c > 0 \tag{9.13}$$

The coefficient of the third term D is identically zero due to the symmetry of the body. However, if D is not identically zero, continuous phase transitions can occur only at isolated points and in such cases Landau has shown that the expansion of G contains no odd order terms, and hence we take $D = 0$. The expansion of G then has the form

$$G(P,\ T,\ \eta) = G_0(P,\ T) + A(P,\ T)\eta^2 + B(P,\ T)\eta^4 \tag{9.14}$$

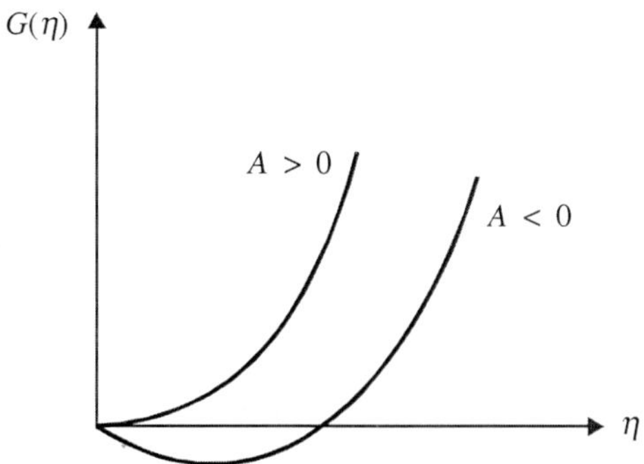

Figure 9.5 Variation of Gibbs potential with order parameter.

where

$$\begin{aligned} &A > 0 \text{ more symmetrical phase} \\ &A < 0 \text{ less symmetrical phase} \\ &B > 0 \end{aligned} \tag{9.15}$$

and the transition points are determined by the condition $A(P, T) = 0$.

In the Landau theory, it is assumed that $A(P, T)$ has no singularity at the transition point and hence can be expanded in the neighbourhood of the transition temperature T_c, in integral powers of $(T - T_c)$. As $(T - T_c)$ is small, we retain only the linear term

$$A(P, T) = a(P)(T - T_c) \tag{9.16}$$

The expansion of the thermodynamic potential G can be now written as

$$G(P, T) = G_0(P, T) + a(P)(T - T_c)\eta^2 + B(P, T_c)\eta^4 \tag{9.17}$$

In the less symmetrical phase, the dependence of η on the temperature near the transition point is determined by equating $\partial G/\partial \eta$ to zero (G is to be minimum). This gives

$$\eta(A + 2B\eta^2) = 0$$

which yields

$$\eta = 0$$

that corresponds with $A < 0$ to a maximum of G

or

$$\eta^2 = -\frac{A}{2B} = a\frac{T_c - T}{2B}$$

which corresponds to the minimum of G.

The two phases depend upon the sign of a. For a being positive or negative, the less symmetrical phase corresponds to $T < T_c$ and $T > T_c$ respectively. As it happens more commonly, the more symmetrical phase is at $T > T_c$ and hence a is positive.

If we neglect higher powers of η, we have for entropy

$$S = -\frac{\partial G}{\partial T} = S_0 - \frac{\partial A}{\partial T}\eta^2 = S_0 + \frac{a^2}{2B}(T - T_c) \tag{9.18}$$

where the temperature derivative of η is zero since $\partial G/\partial \eta = 0$ and $S_0 = -\partial G_0/\partial T$.

In the more symmetrical phase, $\eta = 0$ and we have $S = S_0$; in the less symmetrical phase at $T = T_c$, we have $S = S_0$, which means that the entropy is continuous at the transition point.

For the specific heat $C_P = T(\partial S/\partial T)_P$, we have

$$C_P = C_{P_0} + \frac{a^2}{2B} T_c \tag{9.19}$$

For the more symmetrical phase, $S = S_0$ and $C_P = C_{P0}$. Hence the specific heat is discontinuous at the transition point in the second order phase transition. The other quantities C_V, the thermal expansion coefficient, the compressibility, etc. are also discontinuous. Since volume and entropy are continuous at the transition point, we have

$$\Delta V = 0, \quad \Delta S = 0 \tag{9.20}$$

With pressure and temperature as independent variables, we differentiate the first with respect to temperature

$$\Delta\left(\frac{\partial V}{\partial T}\right)_P + \Delta\left(\frac{\partial V}{\partial P}\right)_T \frac{dP}{dT} = 0 \tag{9.21}$$

and use

$$\left(\frac{\partial S}{\partial P}\right)_T = -\left(\frac{\partial V}{\partial T}\right)_P, \quad \left(\frac{\partial V}{\partial P}\right)_T = -\left(\frac{\partial V}{\partial T}\right)_P \frac{dT}{dP}$$

and

$$\left(\frac{\partial S}{\partial P}\right)_T = -\left(\frac{\partial S}{\partial T}\right)_P \frac{dT}{dP} \quad \text{and} \quad T\left(\frac{\partial S}{\partial T}\right)_P = C_P$$

we get

$$\Delta C_P = T\left(\frac{dP}{dT}\right)^2 \Delta\left(-\frac{\partial V}{\partial P}\right)_T \tag{9.22}$$

or

$$\Delta C_P = T\frac{dP}{dT} \Delta\left(\frac{\partial V}{\partial T}\right)_P \tag{9.23}$$

Equations (9.22) and (9.23) relate the discontinuity of specific heat to the compressibility [$= -(1/V)(\partial V/\partial P)_T$] and the thermal expansion coefficient [$(1/V)(\partial V/\partial T)_P$]. We also note that the discontinuities have the same sign.

With temperature and volume as independent variables, we differentiate the second with respect to temperature, keep in mind that the pressure is unchanged in the transition and use

$$\left(\frac{dS}{dV}\right)_T = \left(\frac{\partial P}{\partial T}\right)_V \quad \text{and} \quad \left(\frac{\partial P}{\partial T}\right)_V = -\left(\frac{\partial P}{\partial V}\right)_T \frac{dV}{dT}$$

to get

$$\Delta C_V = -T\left(\frac{dV}{dT}\right)^2 \Delta\left(-\frac{\partial V}{\partial P}\right)_T^{-1} \tag{9.24}$$

We observe that the compressibility decreases discontinuously in going from less symmetrical to more symmetrical phase.

CRITICAL EXPONENTS

It has already been pointed out that the point of transition of the second kind is a singularity for the thermodynamic functions. This is due to an anomalous increase in the fluctuations of the order parameter, which in turn is due to the flatness of the thermodynamic potential minimum near the transition point. We assume that the change in symmetry is described by only one parameter η.

The mean square fluctuation of the order parameter in a given volume V is

$$\langle \Delta\eta^2 \rangle = \frac{\chi k T_c}{V} \tag{9.25}$$

where χ is susceptibility in a weak external field.

The spatial correlation function is written as

$$C(\vec{r}) = \langle \Delta\eta(\vec{r}_1)\, \Delta\eta(\vec{r}_2) \rangle = \sum_k \langle |\Delta\eta_k|^2 \rangle e^{i\vec{k}\cdot\vec{r}} \tag{9.26}$$

Fourier analysis gives

$$C(r) = \frac{kT_c}{8\pi g r} e^{-r/r_c} \tag{9.27}$$

where r_c is called correlation radius and is a measure of distance over which fluctuation are strongly correlated and g is defined through

$$g\delta_{ik} = g_{ik} \frac{\partial^2 \eta}{\partial x_i \partial x_k} \quad (\delta_{ik} \text{ is Kronecker delta}) \tag{9.28}$$

which is true for the simplest cubic lattice. We have only quoted results here, which we shall use without going into details.

The theories of the second order phase transition are based on plausible assumptions, which are not rigorously proved. Hence they need to be confirmed by experimental results.

Experiments suggest that $\partial C_P/\partial T$ tends to infinity as $T \to T_c$ and in many cases the specific heat itself becomes infinite (Pippard 1956).

Critical exponents or indices are assigned to different parameters in the critical region. The temperature dependence of the specific heat in the fluctuation range is written as

$$C_P \propto |T - T_c|^{-\alpha} \tag{9.29}$$

The exponent α is equal on either side of the transition point, but the proportionality factor is different on the two sides of the transition point. Since $\int C_P\, dT$ is the quantity of heat which is always finite, and if only $\partial C_P/\partial T$ tends to infinity and not C_P, then $-1 < \alpha < 0$. Further, Eq. (9.29) gives only the singular part of the specific heat and hence

$$C_P = C_{P_0} + C_{P_1} |T - T_c|^{-\alpha}$$

The equilibrium value of the order parameter tends to zero in the less symmetrical phase. Accordingly we write

$$\eta \propto \left[-(T - T_c)\right]^{\beta} \tag{9.30}$$

where $\beta > 0$ and refers only to the less symmetrical phase.

For the fluctuation properties of the order parameter, an exponent ν defines the temperature dependence of the correlation radius

$$r_e \propto |T - T_c|^{-\nu} \tag{9.31}$$

where $\nu > 0$.

The decrease of correlation function with increasing distance at $T = T_c$ is described by the exponent ζ as

$$C(r) \propto r^{-(d-2+\zeta)} \tag{9.32}$$

where d is the dimension of space (3 for three-dimensional space). The law is valid for distances $r << r_C$. These laws relate to zero field (no external field) cases.

A group of indices are further used to describe the properties of the body in the fluctuation range when an external h field is present. The variation of susceptibility with temperature for weak fields has an exponent γ as

$$\chi \propto |T - T_c|^{-\gamma} \tag{9.33}$$

where $\gamma > 0$.

For the opposite case of strong fields, we have

$$C_P \propto h^{-\varepsilon} \tag{9.34}$$

$$\eta \propto h^{1/\delta},\ \delta > 0 \tag{9.35}$$

$$r_c \propto h^{-\mu},\ \mu > 0 \tag{9.36}$$

These eight-critical indices are connected by various exact relations called *scaling laws* or *scaling relations*.

The phase transition over a certain temperature range is smoothed out by the application of an external field. The magnitude of its range $(T - T_c)$ is estimated by the condition that the field-induced order parameter $\eta_{\text{ind}}(h)$ and the spontaneous order parameter $\eta_{sp}(T - T_c)$ are of the same order, i.e.

$$\eta_{\text{sp}} \propto (|T - T_c|)^{\beta} \quad \text{and} \quad \eta_{\text{ind}} = \chi h \propto h|T - T_c|^{-\gamma}$$

which gives

$$h \propto |T - T_c|^{\beta+\gamma} \tag{9.37}$$

The thermal part of the thermodynamic potential, which is of the order of $(T - T_c)^2\, C_P$, and the field part of the thermodynamic potential $-V\eta h$ are of the same order. This gives

$$h \propto |T - T_c|^{-2-\alpha-\beta} \tag{9.38}$$

Combining Eqs. (9.37) and (9.38), we get

$$\alpha + 2\beta + \gamma = 2 \tag{9.39}$$

This scaling relation was obtained by Essam and Fisher (1963).

If we combine Eqs. (9.30) and (9.35), we have

$$\eta \propto |T - T_c|^{\beta} \propto h^{1/\delta}$$

and using Eq. (9.37) for h, we get

$$\beta + \gamma = \beta\delta \tag{9.40}$$

the result first obtained by Weedom (1964).

Combining Eqs. (9.29), (9.34) and (9.37), we get

$$\varepsilon(\beta + \gamma) = \alpha \tag{9.41}$$

Similarly combining Eqs. (9.31), (9.36) and (9.37), we get

$$\mu(\beta + \gamma) = \nu \tag{9.42}$$

The integral of the correlation function in the region of space (r_c^d) in which it is strong along with the relation given by Eq. (9.32) gives for the magnitude of the integral in a d-dimensional space

$$\propto r_c^d r_c^{-(d-2+\zeta)} = r_c^{2-\zeta} \propto |T - T_c|^{-\nu(2-\zeta)}$$

in which we have used Eq. (9.31).

The mean square fluctuation from Eq. (9.25) combined with Eq. (9.33) gives

$$\langle \Delta\eta^2 \rangle = \frac{\chi k T_c}{V} \propto |T - T_c|^{-\gamma}$$

Comparison of the above two gives

$$\nu(2 - \zeta) = \gamma \tag{9.43}$$

We have thus defined eight critical indices and five scaling relations between them and hence only three of them can be taken as independent ones.

SCALE INVARIANCE

The scaling relations derived so far have not required any assumption about the nature of fluctuations near the transition point. The spatial distribution of fluctuations involves two characteristic factors: the correlation radius r_c and a region of the body r_0 in which the mean square fluctuation of the order parameter is comparable with its equilibrium value. We must use the condition $r_c >> r_0$ in evaluating the integral of the correlation function. As $(T - T_c) \to 0$, r_0 increases more rapidly than r_c and they become comparable at the boundary of the Landau region. It is assumed that we must have everywhere $r_0 \sim r_c$ so that r_c remains the only dimension characterizing the fluctuations. This is called the *hypothesis of scale invariance* (Kadanoff, 1966, Patashinskii and Pokrovskii, 1966).

The fluctuation in volume is of the order of r_c^d. In formula given by Eq. (9.25), we express χ, r_c and η as powers of $(T - T_c)$ from Eqs. (9.33), (9.31) and (9.30) to get

$$\nu d - \gamma = 2\beta$$

which with the aid of Eq. (9.39) becomes

$$\nu d = 2 - \alpha \tag{9.44}$$

that is known as *Josephson scaling law*. We thus can express all the critical exponents in terms of only two independent ones.

The requirement of scale invariance allows us to derive all the scaling relations in a uniform way. We first make a formal statement of this requirement.

If the scale of all spatial distances change by the same factor $\vec{r} \to \vec{r}/\lambda$, where λ is a constant, then the scale invariance asserts that the scales of measurement of $(T - T_c)$, h and η should be changed in such a way that all the relations in the theory remain unchanged. This means that in the transformations

$$(T - T_c) \to (T - T_c)\lambda^{x_1},\ h \to h\lambda^{x_2},\ \eta \to \eta\lambda^{x_3} \text{ when } \vec{r} \to \vec{r}/\lambda \tag{9.45}$$

the exponents x_1, x_2, x_3, called scaling dimensions, are chosen in such a way that the factor λ does not appear in any of the relations.

The correlation radius must also change by $r_c \to r_c/\lambda$ so that the asymptotic form of correlation function is not altered. Then for the above transformations from Eqs. (9.31) and (9.34), we have

$$x_1 = \frac{1}{\nu},\ x_2 = \frac{1}{\mu} \tag{9.46}$$

The infinitesimal change dG in the thermodynamic potential for an infinitesimal change dh in the field at constant pressure and temperature is given by

$$dG = -V\eta\, dh$$

In the scale transformation, dG is unchanged, i.e.,

$$V\lambda^{-d}\,\eta\lambda^{x_3} dh\lambda^{x_2} = V\eta\, dh$$

This gives

$$x_3 = d - x_2 = d - \frac{1}{\mu} \tag{9.47}$$

Thus, the dimensions x_1, x_2, x_3 are expressed in terms of two critical indices. The requirement of scale invariance of other relations leads to expressions of other critical indices in terms μ and ν. The condition for invariance of equation of state of the system leads to

$$(\mu\delta - 1)\delta = 1,\quad \mu\left(\frac{2}{\nu} - \delta\right) = \varepsilon \tag{9.48}$$

It is easily verified that they follow from the relations already derived.

EXERCISES

1. What is a phase transition of second order? How is it different from ordinary liquid-gas transition?
2. Discuss Landau theory of second order phase transition. How is the discontinuity in the specific heat related to the discontinuity in isothermal compressibility and the discontinuity in thermal expansion coefficient?
3. What are critical indices? Develop the different scaling relations connecting the critical indices.
4. For $(T - T_c) \to 0$, if we take more accurately

$$C_P = T_c \frac{\partial P_c}{dT}\left(\frac{\partial V}{\partial T}\right)_P + a; \quad \left(\frac{\partial V}{\partial T}\right)_P = -\left(\frac{\partial V}{\partial P}\right)_T \frac{dP_c}{dT} + \frac{b}{T_c}\frac{dT_c}{dP}$$

where a and b are constants; and if $C_P \propto (T - T_c)^{-\alpha}$ with $\alpha > 0$, show that

$$\frac{\partial C_V}{\partial T} \propto (T - T_e)^{-(1-\alpha)}$$

Hint: $C_P - C_V = -T\left[\left(\frac{\partial V}{\partial T}\right)_P\right]^2\left(\frac{\partial P}{\partial V}\right)_T$

Substitution for $(\partial V/\partial T)_P$ and $(\partial P/\partial V)_T$ gives

$$C_V = a - b - b^2/C_P$$

and

$$\frac{\partial C_V}{\partial T} = \frac{b^2}{C_P^2}\frac{\partial C_P}{\partial T}$$

Thus

$$\frac{\partial C_V}{\partial T} \propto (T - T_e)^{-(1-\alpha)}$$

5. Discuss scale invariance. How does it allow us to obtain the scaling relations in a uniform way?

10 The Ising Model

COOPERATIVE PHENOMENON

Let us consider the phenomenon of ferromagnetism. In some metals like iron and nickel, a finite fraction of spins of the atoms are spontaneously aligned in the same direction, without any external field below a certain temperature T_c called the *Curie temperature*. This gives rise to a macroscopic magnetic field. As the temperature increases, some of the aligned spins flip over due to the thermal energy. The regular ordering is destroyed rapidly at the Curie temperature and above it; the spins are oriented at random, producing no net magnetic field. This transition from non-ferromagnetic state to ferromagnetic state is a phase transition of the second kind and involves the symmetry of spins. This is similar to the order-disorder transition in the case of an alloy. However, in the case of an alloy, the number of constituent atoms is fixed whereas in the non-ferromagnetic to ferromagnetic transition, the spins can transform freely into one another. Transitions of this kind come under a group of *cooperative phenomena*, in which certain subsystems like spins cooperate due to exchange interactions to form units below a certain critical point.

The Heisenberg model in which the exchange interaction is $\sum_{\langle i,j\rangle} \epsilon_{ij}\vec{s}_i\cdot\vec{s}_j$ (where $\vec{s}_i$ is spin variable), in which the sum $\langle i, j\rangle$ extends over only nearest-neighbour pairs, has proved to be difficult problem for complete solution. A simpler method in which the product $\vec{s}_i\cdot\vec{s}_j = s_{ix}s_{jx} + s_{iy}s_{jy} + s_{iz}s_{jz}$ is replaced by $s_{iz}s_{jz} \equiv s_is_j$ (from now onwards) was introduced by Ising in 1925. The one-dimensional case was solved by Ising himself, the two-dimensional case was solved by Onsager in 1944, and the three-dimensional case is yet to be solved completely. However, many details of the statistical ordering near the transition are obtained by refined mathematical techniques.

FORMULATION OF ISING MODEL

Ising model is an attempt to simulate the structure of a ferromagnetic substance. In this model, the system is considered to be an array of N fixed lattice sites forming an n-dimensional (n = 1, 2, 3) periodic lattice. Let us associate a spin variable s_i (i = 1, 2,..., N) with each lattice site which is either +1 (spin up) or –1 (spin down). A set of numbers $\{s_i\}$ will then specify a configuration of the whole system. The energy of the system in the configuration specified by $\{s_i\}$ in Ising model in a magnetic field $\vec{B}$ is

$$H_I\{s_i\} = -\sum_{\langle i,j\rangle} \in_{ij} s_i s_j - \mu B \sum_{i=1}^{N} s_i \tag{10.1}$$

where I stands for Ising model and $\langle i, j\rangle$ denotes the nearest-neighbour pairs. There is no difference between $\langle i, j\rangle$ and $\langle j, i\rangle$. The sum $\langle i, j\rangle$ will have $\gamma N/2$ terms, where γ is the number of nearest-neigbhours, and depends on the geometry of the lattice. γ is given by

$$\gamma = \begin{cases} 4 \text{ (two-dimensional square lattice)} \\ 6 \text{ (three-dimensional cubic lattice)} \\ 8 \text{ (three-dimensional body centred cubic lattice)} \end{cases}$$

We assume that the interactions are isotropic, i.e., $\in_{ij} = \in$. The Hamiltonian then becomes

$$H_I\{s_i\} = -\in \sum_{\langle i,j\rangle} s_i s_j - \mu B \sum_{i=1}^{N} s_i \tag{10.2}$$

The case $\in > 0$ corresponds to ferromagnetism and the case $\in < 0$ to antiferromagnetism.

The partition function is

$$Q_I(B,T) = \sum_{s_1} \sum_{s_2} \cdots \sum_{s_N} e^{-H_I\{s_i\}/kT} \tag{10.3}$$

The Helmholtz free energy and subsequently the energy, the specific heat and the magnetization are given by

$$A_I(B,T) = -kT \ln Q_I(B,T) \tag{10.4}$$

$$E_I(B,T) = -kT^2 \frac{\partial}{\partial T}\left(\frac{A_I}{kT}\right) \tag{10.5}$$

$$C_I(B,T) = \frac{\partial E_I}{\partial T} \tag{10.6}$$

$$M_I(B,T) = -\frac{\partial}{\partial B}\left(\frac{A_I}{kT}\right) = \mu\left\langle \sum_{i=1}^{N} s_i \right\rangle \tag{10.7}$$

where $\langle\rangle$ denotes ensemble average. The quantity $M_I(0, T)$ is spontaneous magnetization and is non-zero for ferromagnetic system.

Let N_+ be the total number of spin-up and N_- be the total number of spin-down. Then, $N_+ = N - N_-$ and, the number of pairs be N_{++} (two spin up pair), N_{--} (two spin-down pair) and N_{+-} (one spin-up, one spin-down pair, no distinction between N_{+-} and N_{-+}). To find a relation between them, let us choose a square lattice. Choose a lattice site with spin-up and draw a line to all its nearest neighbours. There will be γ lines. Further, draw a line between spin-up pairs. Then there will be two lines between spin-up pairs, one line between spin-up and spin-down pair and no line between two spin-down pairs This gives the relation

$$\left.\begin{aligned} \gamma N_+ &= 2N_{++} + N_{+-} \\ \gamma N_- &= 2N_{--} + N_{+-} \\ N &= N_+ + N_- \end{aligned}\right\} \tag{10.8}$$

This reduces to

$$\left.\begin{aligned} N_{+-} &= \gamma N_+ - 2N_{++} \\ N_{--} &= \frac{\gamma}{2} N + N_{++} - \gamma N_+ \end{aligned}\right\} \tag{10.9}$$

Also, we have

$$\left.\begin{aligned} \sum_{\langle i,j\rangle} s_i s_j &= N_{++} + N_{--} - N_{+-} = 4N_{++} - 2\gamma N_+ + \frac{\gamma}{2} N \\ \sum_{i=1}^{N} s_i &= N_+ - N_- = 2N_+ - N \end{aligned}\right\} \tag{10.10}$$

With the aid of Eqs. (10.10) and (10.9), we have

$$H_I\{s_i\} \equiv H_I(N_+, N_{++}) = -4\in N_{++} + 2(\in\gamma - \mu B)N_+ - \left(\frac{1}{2}\gamma\in - \mu B\right)N \tag{10.11}$$

Thus, the Hamiltonian of a state depends upon only two numbers, N_{++} and N_+, which express certain features of the spin distribution.

BRAGG–WILLIAMS APPROXIMATION

Let the spin distribution be random for a given value of N_+ and N_{++}. N_{++} is the number of nearest neighbour pairs with spin-up out of a total of $\gamma N/2$ pairs. The quantity $N_{++}/(\gamma N/2)$ is then the fraction of nearest neighbour pairs having spin-up, and is a measure of the local correlation of spin. The fraction N_+/N needs no correlation between nearest neighbours, and is a measure of lattice fraction of all the spins-up.

We define a short-range order parameter ω and a long-range order parameter L by

$$\frac{N_{++}}{1/2\,\gamma N} = \frac{1}{2}(\omega + 1) \tag{10.12}$$

$$\frac{N_+}{N} = \frac{1}{2}(L + 1) \tag{10.13}$$

Obviously, we have

$$\left.\begin{aligned} -1 \le L \le 1 \\ -1 \le \omega \le 1 \end{aligned}\right\} \tag{10.14}$$

Use of Eqs. (10.12) and (10.13) in Eq. (10.10) gives

$$\sum_{\langle i,j\rangle} s_i s_j = \frac{1}{2}\gamma N(2\omega - 2L + 1) \tag{10.15}$$

$$\sum_{i=1}^{N} s_i = NL \tag{10.16}$$

The ensemble average of the long-range order parameter is a measure of the magnetization per particle. The Hamiltonian becomes

$$H_I(L,\omega) = -\frac{N}{2}\in \gamma(2\omega - 2L + 1) - N\mu BL \tag{10.17}$$

The Bragg–Williams approximation states that *there is no short-range order apart from that which follows from long-range order*. Mathematically, this means putting

$$\frac{N_{++}}{\frac{1}{2}\gamma N} \simeq \left(\frac{N_+}{N}\right)^2 \tag{10.18}$$

This gives

$$\omega \simeq \frac{1}{2}(L+1)^2 - 1 \tag{10.19}$$

In Bragg–Williams approximation, Eq. (10.17) reduces to

$$H_I(L) \simeq -\frac{N}{2}\in \gamma L^2 - N\mu BL \tag{10.20}$$

and the partition function given by Eq. (10.3) now becomes

$$Q_I(B,T) = \sum_{\{s_i\}} e^{N(1/2\in\gamma L^2 + \mu BL)/kT} \tag{10.21}$$

The summand depends only on L, though the sum is over all sets $\{s_i\}$. Therefore, we find the number of sets $\{s_i\}$ that share same L.

The number of arrangements of spins-up over N sites is clearly the number of ways to pick N_+ things out of N, i.e.

$$\frac{N!}{N_+!\,(N-N_+)!} = \frac{N!}{[(1/2)N(1+L)]!\,[(1/2)N(1-L)]!}$$

Hence we have

$$Q_I(B,T) = \sum_{L=-1}^{L=1} \frac{N!}{[(1/2)N(1+L)]!\,[(1/2)N(1-L)]!}\, e^{N[(1/2)\in\gamma L^2 + \mu BL]/kT} \tag{10.22}$$

When $N \to \infty$, the logarithm of the partition function is equal to the logarithm of the largest term in the summand. Using Stirling approximation

$$\ln N! = N \ln N - N$$

we obtain

$$\ln Q_I(B, T) = \frac{N[(1/2)\in\gamma\bar{L}^2 + \mu B\bar{L}]}{kT} - \frac{N(1+\bar{L})}{2}\ln\frac{(1+\bar{L})}{2} - \frac{N(1-\bar{L})}{2}\ln\frac{(1-\bar{L})}{2} \quad (10.23)$$

where $\bar{L}$ is the value of L that makes the summand maximum. This is obtained by putting

$$\frac{\partial \ln Q_I}{\partial L} = 0$$

the resulting value $\bar{L}$, is given by

$$\frac{1}{2}\ln\frac{1+\bar{L}}{1-\bar{L}} = \frac{\mu B + \in\gamma\bar{L}}{kT} \quad (10.24)$$

The left-hand side of Eq. (10.24) is $\tanh^{-1}\bar{L}$. This gives

$$\bar{L} = \tanh\left(\frac{\mu B + \in\gamma\bar{L}}{kT}\right) \quad (10.25)$$

For spontaneous magnetization ($B = 0$), Eq. (10.25) reduces to

$$\bar{L} = \tanh\left(\frac{\in\gamma\bar{L}}{kT}\right) \quad (10.26)$$

The equation can be solved graphically. The solutions are the intersection of graph tanh $(\in\gamma\bar{L}/kT)$ against $\bar{L}$ and the straight line of slope $\pi/4$ (Fig. 10.1).

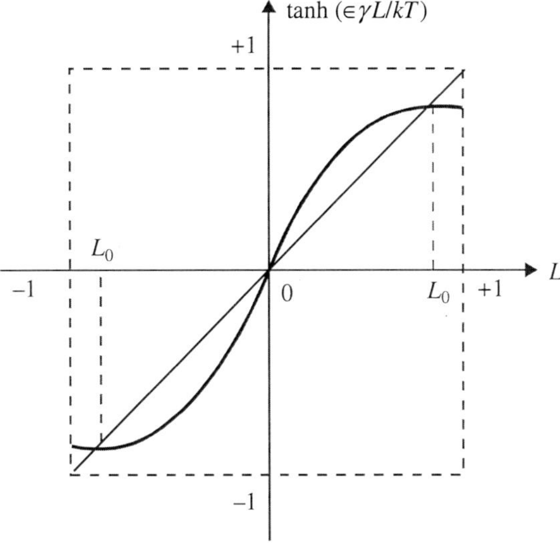

Figure 10.1 A graphical solution of Eq. (10.26).

The solutions are

$$\bar{L} = \begin{cases} L_0 \\ 0 & \text{for } \dfrac{\in\gamma}{kT} > 1 \\ -L_0 \end{cases} \tag{10.27}$$

$$\bar{L} = 0 \quad \text{for } \frac{\in\gamma}{kT} < 1$$

where L_0 is the root of Eq. (10.26).

The solution $\bar{L} = 0$ is rejected as it corresponds to a minimum (easily seen by substituting the value in Eq. (10.23)). For $\in > 0$, there exists a critical temperature T_c given by

$$kT_c = \in\gamma \tag{10.28}$$

such that

$$\bar{L} = \begin{cases} 0 & T > T_c \\ \pm L_0 & T < T_c \end{cases} \tag{10.29}$$

Since $\bar{L}$ is the magnetization per particle, we find that the system is ferromagnetic for $T < T_c$ and has no magnetization for $T > T_c$. T_c is the Curie temperature of the system.

In general, L_0 has to be computed numerically from

$$L_0 = \tanh\left(\frac{\in\gamma L_0}{kT}\right) \tag{10.30}$$

but we can examine the values for $T \to 0$ K and $T \to T_c$.

For $T \to 0$ K, we have $L_0 \to 1$ and we can write Eq. (10.30) using Eq. (10.28) as

$$L_0 = \frac{1 - e^{-2L_0T_c/T}}{1 + e^{-2L_0T_c/T}} \simeq 1 - 2e^{-2T_c/T} \tag{10.31}$$

For $T \to T_c$, we have L_0 small and we can write Eq. (10.30) using Eq. (10.28) as

$$L_0 \simeq \frac{L_0 T_c}{T} - \frac{1}{3}\left(\frac{L_0 T_c}{T}\right)^3$$

which gives

$$L_0 \simeq \sqrt{3\left(1 - \frac{T}{T_c}\right)} \tag{10.32}$$

The thermodynamic functions can be written as

$$A_I(0,T) = \begin{cases} 0 & \text{for } T > T_c \\ \dfrac{N\in\gamma}{2} + \dfrac{NkT}{2}\ln\dfrac{1 - L_0^2}{4} & \text{for } T < T_c \end{cases} \tag{10.33}$$

$$M_I(0,T)=\begin{cases}0 & \text{for } T>T_c\\ \mu NL_0 & \text{for } T<T_c\end{cases} \tag{10.34}$$

$$E_I(0,T)=\begin{cases}0 & \text{for } T>T_c\\ -\dfrac{N\in\gamma}{2}L_0^2 & \text{for } T<T_c\end{cases} \tag{10.35}$$

$$C_I(0,T)=\begin{cases}0 & \text{for } T>T_c\\ -\dfrac{N\in\gamma}{2}\dfrac{dL_0^2}{dT} & \text{for } T<T_c\end{cases} \tag{10.36}$$

If we use Eq. (10.32), we get

$$C_I(0,T)=\frac{3}{2}Nk \tag{10.37}$$

A graph of $L_0 - kT$ and $C_I/Nk - kT$ is shown in Fig. 10.2 and Fig. 10.3.

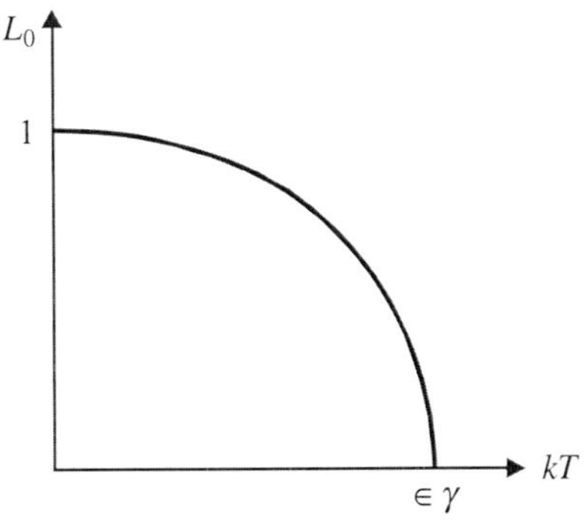

Figure 10.2 Spontaneous magnetization.

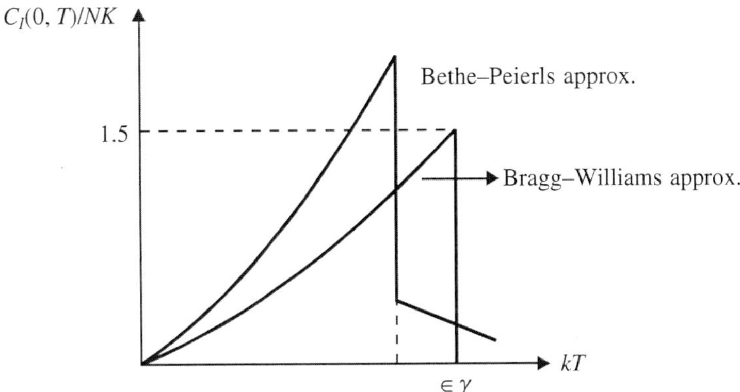

Figure 10.3 Specific heat.

We see that the thermodynamic quantities vanish for $T > T_c$. This is because both the short-range order and the long-range order vanish in that region in Bragg–Williams approximation.

Bethe-Peierls approximation is an improvement over the Bragg–Williams approximation, which includes specific short-range order in the calculations. The specific heat in the Bethe–Peierls approximation does not vanish for $T > T_c$ and gives a better agreement with experiments than the Bragg–Williams approximation.

ONE-DIMENSIONAL ISING MODEL

The one-dimensional Ising model is a chain of N spins, each spin interacting with its nearest neighbour and with an external magnetic field $\vec{B}$. The Hamiltonian is given by

$$H_I = -\epsilon \sum_{i=1}^{N} s_i s_{i+1} - \mu B \sum_{i=1}^{N} s_i \tag{10.38}$$

We use periodic boundary condition

$$s_{N+1} = s_N \tag{10.39}$$

This makes the topology of chain that of a ring, as shown in Fig. 10.4.

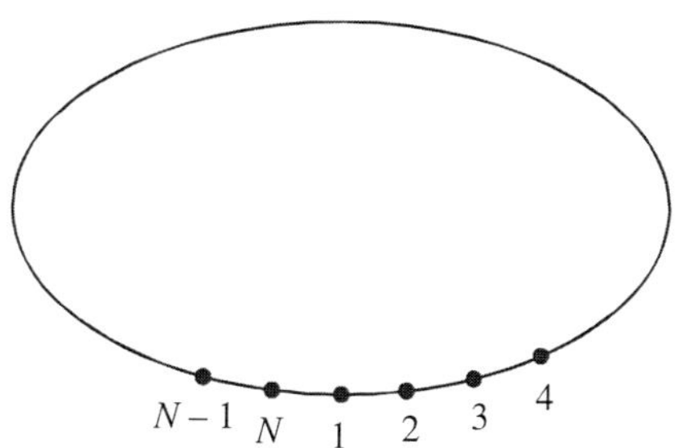

Figure 10.4 One-dimensional Ising lattice topology.

The partition function becomes

$$Q_I(B,T) = \sum_{s_i} \cdots \sum_{s_N} e^{\sum_{i=1}^{N} (\epsilon s_i s_{i+1} + \mu B s_i)/kT} \tag{10.40}$$

where each s_i independently assumes the value ± 1. This can be written using Eq. (10.39) as

$$Q_I(B,T) = \sum_{s_i} \cdots \sum_{s_N} e^{\sum_{i=1}^{N} [\epsilon s_i s_{i+1} + (1/2)\mu B (s_i + s_{i+1})]/kT} \tag{10.41}$$

Let us define a 2×2 matrix $[P]$, whose matrix elements are

$$\langle s|P|s'\rangle = e^{[\epsilon s s' + (1/2)\mu B(s + s')]/kT}$$

We then have

$$\left.\begin{array}{l}\langle +1|P|+1\rangle = e^{(\in+\mu B)/kT} \\ \langle -1|P|-1\rangle = e^{(\in-\mu B)/kT} \\ \langle +1|P|-1\rangle = \langle -1|P|+1\rangle = e^{-\in/kT}\end{array}\right\} \tag{10.42}$$

The explicit form of the matrix is

$$[P] = \begin{bmatrix} e^{(\in+\mu B)/kT} & e^{-\in/kT} \\ e^{-\in/kT} & e^{(\in+\mu B)/kT} \end{bmatrix} \tag{10.43}$$

The partition function may be rewritten as

$$Q_I(B,T) = \sum_{s_1}\sum_{s_2}\cdots\sum_{s_N} \langle s_1|P|s_2\rangle\langle s_2|P|s_3\rangle\cdots\langle s_N|P|s_1\rangle$$

which on account of periodic condition given by Eq. (10.39) becomes trace of the Nth power of the matrix $[P]$, i.e.

$$Q_I(B,T) = Tr.\, P^N = \lambda_+^N + \lambda_-^N \tag{10.44}$$

where λ_+ and λ_- are the two eigenvalues of the matrix $[P]$ with $\lambda_+ \geq \lambda_-$. The trace is evaluated by bringing the matrix in a diagonal form.

The two eigenvalues can be calculated as

$$\lambda_\pm = e^{\in/kT}\left[\cosh\,(\mu B/kT) \pm \sqrt{\cosh^2(\mu B/kT) - 2e^{-2\in/kT}\sinh(2\in/kT)}\,\right] \tag{10.45}$$

From Eq. (10.44) we have

$$\ln Q_N = N\left\{\ln\lambda_+ + \ln\left[1 + \left(\frac{\lambda_-}{\lambda_+}\right)^N\right]\right\} \underset{N\to\infty}{\to} N\ln\lambda_+ \tag{10.46}$$

We thus find that only the larger eigenvalue λ_+ is relevant.

The Helmholtz free energy with the aid of Eq. (10.4) is

$$A_I(B,T) = -N\in - NkT\ln[\cosh\,(\mu B/kT) + \sqrt{\cosh^2(\mu B/kT) - 2e^{-2\in/kT}\sinh^2(2\in/kT)}] \tag{10.47}$$

and the magnetization is

$$M_I(B,T) = \frac{N\sinh\,(\mu B/kT)}{\sqrt{\cosh^2(\mu B/kT) - 2e^{-2\in/kT}\sinh^2(2\in/kT)}} \tag{10.48}$$

A graph of $M_I(B, T)/N$ against B is shown in Fig. 10.5.

We immediately see that there is no spontaneous magnetization $\{M_I(0, T) = 0\}$ and hence no transition takes place. Thus, the one-dimensional Ising model cannot be ferromagnetic. This happens because in the one-dimensional model, the tendency for alignment loses out due to there not being enough nearest neighbours.

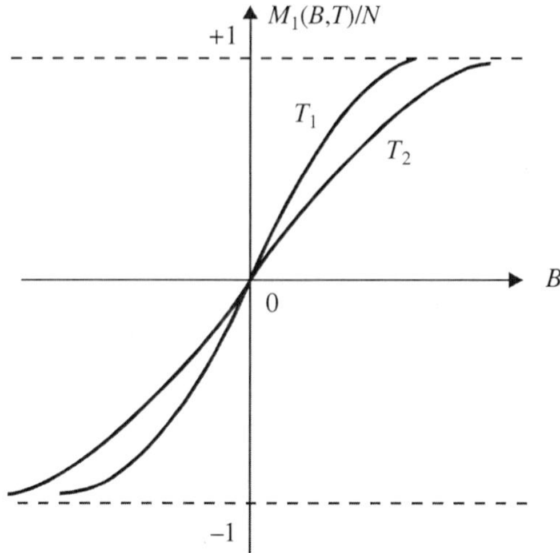

Figure 10.5 Magnetization in one-dimensional Ising model.

TWO-DIMENSIONAL ISING MODEL

Let us take a square lattice of n rows and n columns. Then, $N = n^2$. Let $\xi_\alpha(\alpha = 1, 2, \ldots, n)$ denote the collection of all the spin coordinates of αth row, i.e.,

$$\xi_\alpha \equiv \{s_1, s_2, \ldots, s_n\} \tag{10.49}$$

Let us now imagine the lattice to be enlarged by one row and one column in such way that the configuration of $(n + 1)$th row and column is identical with that of the first row and column respectively. This leads to toroidal boundary condition

$$\xi_{n+1} \equiv \xi_n \tag{10.50}$$

Because of nearest negighbour interactions only, the αth row interacts with $(\alpha + 1)$th and $(\alpha - 1)$th rows only. If we write $\in(\xi_\alpha, \xi_{\alpha+1})$ and $\in(\xi_\alpha)$ as the interaction energy between the αth and $(\alpha + 1)$th row and the interaction energy within the αth row plus the interaction energy with an external magnetic field, then

$$\left.\begin{aligned} \in(\xi, \xi') &= -\in \sum_{i=1}^{n} s_i s_i' \\ \text{and} \qquad \in(\xi) &= -\in \sum_{i=1}^{n} s_i s_{i+1} - \mu B \sum_{i=1}^{n} s_i \end{aligned}\right\} \tag{10.51}$$

where the collection of spins in two neighbouring rows are

$$\xi \equiv \{s_1, s_2, \ldots, s_n\} \quad \text{and} \quad \xi' \equiv \{s_1' \; s_2', \ldots, s_n'\} \tag{10.52}$$

The toroidal boundary condition implies that

$$s_{n+1} = s_1 \tag{10.53}$$

This gives for the Hamiltonian

$$H_I(B,T) = \sum_{\alpha=1}^{n} [\in(\xi_\alpha, \xi_{\alpha+1}) + \in(\xi_\alpha)] \tag{10.54}$$

The partition function is

$$Q_I(B,T) = \sum_{\xi_1} \cdots \sum_{\xi_n} e^{-\sum_{\alpha=1}^{n} [\in(\xi_\alpha, \xi_{\alpha+1}) + \in(\xi_\alpha)]/kT} \tag{10.55}$$

We define a $2^n \times 2^n$ matrix $[P]$ whose matrix elements are

$$\langle \xi | P | \xi' \rangle = e^{-[\in(\xi, \xi') + \in(\xi)]/kT} \tag{10.56}$$

Then,

$$Q_I(B,T) = TrP^n \tag{10.57}$$

The trace is evaluated by bringing the matrix into its diagonal form, where $\lambda_1, \lambda_2, \ldots, \lambda_{2^n}$, are the 2^n eigenvalues of $[P]$, we then have

$$Q_I(B,T) = \sum_{\alpha=1}^{2^n} (\lambda_\alpha)^n \tag{10.58}$$

We expect all the eigenvalues to be positive, and if $\lambda_{\max}$ is the largest eigenvalue, it can be shown that

$$\underset{N\to\infty}{\text{Lt}} \frac{1}{N} \ln Q_I = \underset{n\to\infty}{\text{Lt}} \frac{1}{n} \ln \lambda_{\max} \tag{10.59}$$

Thus the problem reduces to find the description of an explicit representation for $[P]$.

The evaluation of the explicit representation for $[P]$, its eigenvalues and then the evaluation of the partition function (and subsequently the thermodynamic functions) is extremely complicated. Onsager, *L.* (Phys. Rev. **65**, 117, 1944) solved the problem. The matrix formulation presented here is due to Kaufmann, B. (Phys. Rev. **76**, 1232, 1949). We quote here the results only.

The Helmholtz free energy per spin $a_I(0, T)$ is

$$\beta a_I(0,T) = -\ln(2\cosh 2\beta\in) - \frac{1}{2\pi}\int_0^{\pi} d\phi \ln \frac{1}{2}\left(1 + \sqrt{1 - K^2 \sin^2 \phi}\right) \tag{10.60}$$

where

$$\phi = \beta\in, \quad \in > 0, \quad \beta = \frac{1}{kT}$$

and

$$K = \frac{2}{\cosh 2\phi \coth 2\phi}$$

All thermodynamic functions have a singularity of some kind at $T = T_c$, where T_c is given by

$$2\tanh^2\left(\frac{2\in}{kT_c}\right) = 1$$

This gives

$$kT_c = 2.269\in \tag{10.61}$$

The specific heat in the neighbourhood of T_c for zero magnetic field is given by

$$\frac{1}{k}C_I(0,T) \simeq \frac{2}{\pi}\left(\frac{2\in}{kT_c}\right)^2\left[-\ln\left|1-\frac{T}{T_c}\right|+\ln\left(\frac{kT_c}{2\in}\right)-\left(1+\frac{\pi}{4}\right)\right] \tag{10.62}$$

which tends to infinity as $T \to T_c$. A graph of specific heat is shown in Fig. 10.6.

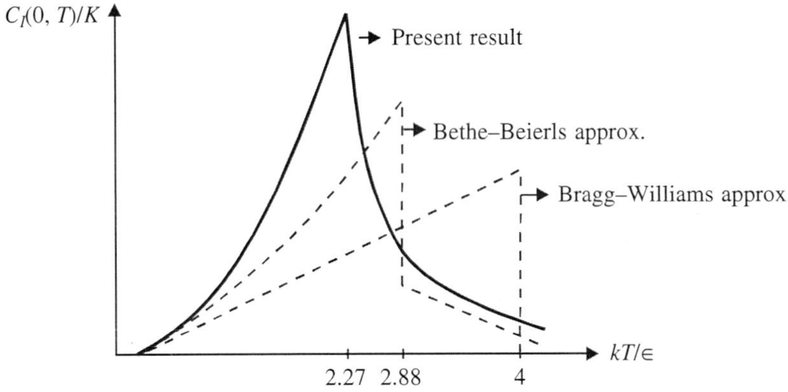

Figure 10.6 Specific heat in two-dimensional Ising model.

To calculate spontaneous magnetization, one has to evaluate the derivative of free energy with respect to B at B = 0. It was done by Yang, C.N. (Phys. Rev. **85**, 809, 1952) and spontaneous magnetization per spin is

$$m_I(0,T) = \begin{cases} 0 & \text{for } T > T_c \\ \dfrac{(1+z^2)^{1/4}(1-6z^2+z^4)^{1/8}}{\sqrt{1-z^2}} & \text{for } T < T_c \end{cases} \tag{10.63}$$

where magnetic moment per spin has been taken to be unity and $z = e^{-2\in/kT}$. The transition temperature is given by $z_c = \sqrt{2}-1$. A graph of spontaneous magnetization is shown in Fig. 10.7.

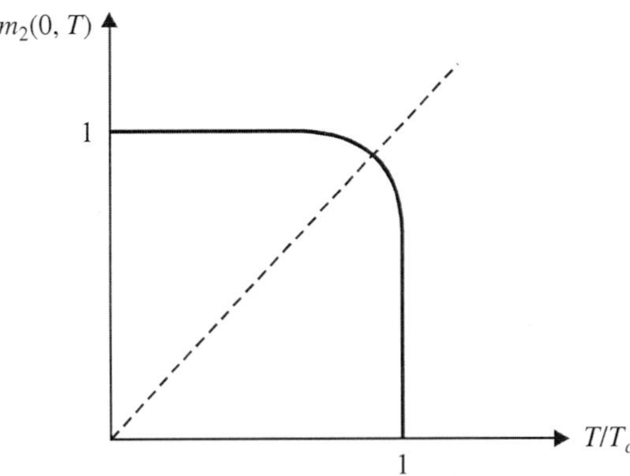

Figure 10.7 Spontaneous magnetization in two-dimensional Ising model.

We thus see that the solution of two-dimensional Ising model describes a phase transition of the second kind.

THREE-DIMENSIONAL ISING MODEL

We have mentioned earlier that the three-dimensional Ising problem has not been completely solved but notable advance has been made towards it. The procedures involve tedious mathematical calculations of successive approximations. Series expansions of partition function at $T \sim 0$ K (low temperature side) and $1/T \to 0$ (high temperature side) are obtained. The results are improved by evaluating the higher order terms. The high temperature expansion is put in a form in which the coefficients are all positive. The radius of convergence gives the location of T_c and the behaviour in the neighbourhood of T_c is obtained by extrapolation. In the low temperature expansions, terms with both positive and negative signs occur and Pade extrapolation technique is used for the location of T_c and the behaviour in its neighbourhood. For a simple cubic lattice, the critical temperature T_c corresponds to $kT_c/\gamma\epsilon = 0.375$. The susceptibility for $T > T_c$ has a temperature variation of the form $(T - T_c)^{-\alpha}$ with $\alpha \simeq 5/4$ just above T_c. The spontaneous magnetization in the ferromagnetic phase tends to zero as T_c is approached with an asymptotic form $(T_c - T)^{\beta}$, with $\beta \simeq 5/16$. These compare well with the experimental values $\alpha \simeq 1.2$ to 1.4 and $\beta \simeq 0.31$ to 0.35. The specific heat reaches a very large value near T_c and persists on the high temperature side for some range of temperature. More terms in the series expansions need to be computed to give unambiguous precise behaviour near T_c. So far, there is good qualitative and also some quantitative agreement with the experimental results.

The Ising model gives a good account of many other features, like residual entropy remaining above T_c, the configurational energy at T_c, etc. Also, the liquid-gas critical point, the order–disorder transition in binary alloys and adsorption of gases on localized sites are fairly well described by the Ising model. The Ising model has received some good attention because it is the simplest non-trivial model to describe semiquantitatively many of the features observed in a broad range of cooperative transitions.

EXERCISES

1. What is Ising model? Discuss Bragg–Williams approximation for the solution of the problem and its shortcomings.
2. Discuss the behaviour of the specific heat near T_c in the Bragg–Williams approximation.
3. Show that the one-dimensional Ising model does not explain the spontaneous magnetization. How does the solution of the two-dimensional Ising model overcome this difficulty?

11 Liquid Helium

THE LAMBDA TRANSITION

Helium is a remarkable substance. Historically, it was first discovered in the sun and then on the earth. In 1908, Kammerlingh Onnes succeeded in liquefying helium (critical temperature 5.2 K). The liquid-gas vapour pressure curve extrapolates to the origin, with no sign of the triple point, i.e., it remains liquid to almost absolute zero. The solid phase of helium comes into being only under an external pressure of 25 atm. There are two isotopes of helium of mass three and four (in atomic units) designated He^3 and He^4. At temperatures of the order of 1 K or 2 K, the de Broglie wavelength of the helium atoms is comparable to the interatomic distance and hence has quantum properties. He^4 contains 2 protons, 2 neutrons and 2 electrons, total spin is zero and hence obeys Bose–Einstein statistics. The isotope He^3 has half integral spin and obeys Fermi–Dirac statistics. The liquid He^4 shows a dramatic change in properties at the temperature 2.18 K called T_λ. For $T < T_\lambda$, the viscosity of liquid He^4 tends to zero (under certain conditions). The phenomenon is called *superfluidity* (Kapitza, 1938). This separates the liquid He^4 into two phases, called He I and He II. The transition from He I to He II is not accompanied by a latent heat. The specific heat along the vapour pressure curve becomes logarithmically infinite as the point of the transition is approached from either side. The shape of the specific heat curve near the transition point has the shape of the letter lambda. This has given rise to the name lambda transition (λ transition). It is a phase transition of the second kind. The two phases of the liquid He^4 and the specific heat curve is shown in Fig. 11.1 and Fig. 11.2.

It should be noted that liquid He^4, which are Bose particles, exhibit superfluidity and λ transition whereas liquid He^3, which are Fermi particles, do not exhibit superfluidity in the above temperature range (in later years it has become clear that in a Fermi liquid of He^3 at sufficiently low temperatures of the order of 0.001 K, pairing should occur, i.e. *bosons* will

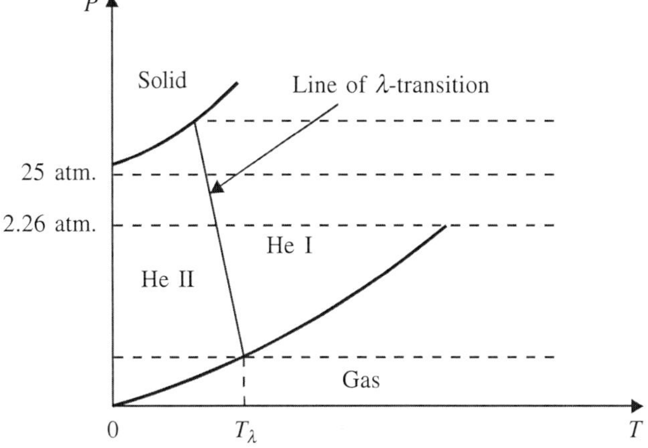

Figure 11.1 Phases of He^4.

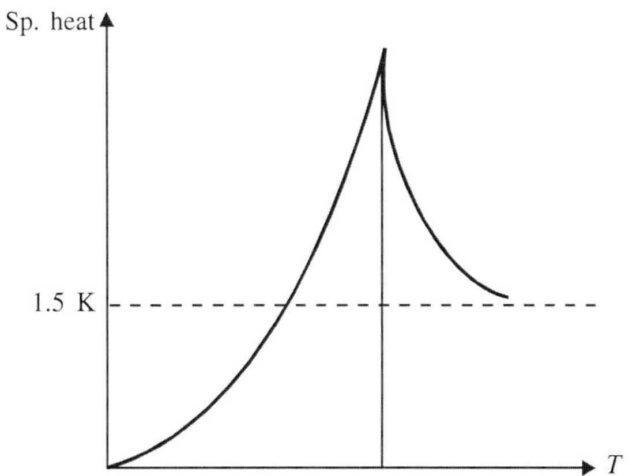

Figure 11.2 Specific heat of liquid He^4 (experimental) along the vapour pressure curve.

be formed, leading to the occurrence of superfluidity). For this reason, it is tempting to assume that the λ transition is the Bose–Einstein condensation modified by the interparticle interactions. An ideal Bose fluid with mass and density identical to liquid He^4 will undergo a Bose–Einstein condensation at 3.14 K, which is of the same order as T_λ. This fact, along with the similarity in the specific heat curve of He^4 with that of an ideal Bose fluid, led London (1938) to suggest that the lambda transition of He^4 is a form of Bose–Einstein condensation.

The weak interparticle interactions between He atoms (Helium is a noble gas) and the small mass of the helium atoms give rise to the fluidity of helium. The other noble elements because of their greater mass and hydrogen, though having lighter molecules, have strong inter-molecular interactions and solidify at comparatively higher temperatures. The above two reasons lead to a large zero-point motion of the He atoms and localization of the atoms at well-defined lattice sites is not possible. This could be easily understood by examining the potential energy (which has the form shown below) between two helium atoms.

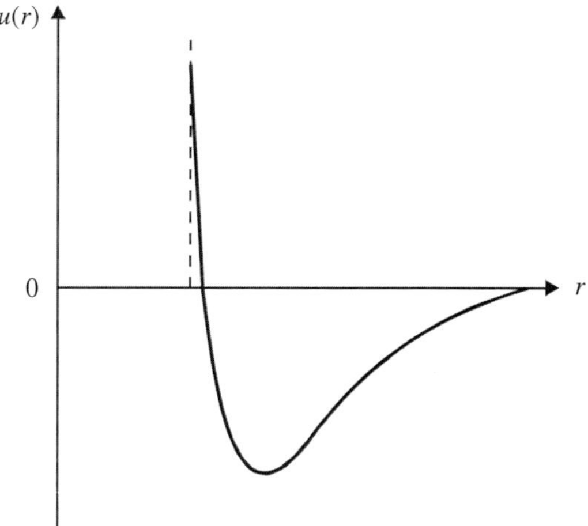

Figure 11.3 Pair potential between two helium atoms.

The depth of potential well is of the order of 10 K for the Lennard–Jones potential. The average interparticle distance is of order of 4.44 Å. We can easily see that even a small (compared with the interparticle distance) uncertainty of the order of 0.5 Å in the localization of the atom gives rise to an uncertainty in the energy of the same order as the depth of the potential well, viz.

$$\Delta E = \frac{(\Delta p)^2}{2m} = \frac{(\hbar/\Delta x)^2}{2m} \sim 10 \text{ K} \quad \text{for } \Delta x \sim 0.5 \text{ Å}$$

This makes the localization of atoms impossible and Helium remains a liquid almost down to absolute zero.

The explicit calculation of the partition function of the liquid helium has not been so far possible (the interparticle interaction not being exactly known). Hence the connection between Bose–Einstein condensation and the λ transition remains only a plausible conjecture.

The Bose–Einstein condensation does take place in liquid He^4, as has been shown by Penrose and Onsager (Phys. Rev. **104**, 576, 1956). Making use of quantized fields formalism, the statistical average of the number of particles with momentum $\vec{p}$ is

$$\langle n_p \rangle = Tr(\hat{w}\,\hat{a}_p^{\dagger}\,\hat{a}_p) = \frac{Tr(e^{-\beta\hat{H}}\,\hat{a}_p^{\dagger}\hat{a}_p)}{Tre^{-\beta\hat{H}}} \tag{11.1}$$

where $\hat{w}$ is the probability density operator of the system in the canonical ensemble and $\hat{a}_p^{\dagger}$ and $\hat{a}_p$ are the creation and annihilation operator for a free particle state of momentum $\vec{p}$. The criterion for Bose–Einstein condensation is

$$\underset{N \to 0}{\text{Lt}} \frac{\langle n_0 \rangle}{N} > 0 \tag{11.2}$$

The one particle density matrix in coordinate representation is given by

$$\rho_1(\vec{r}_1, \vec{r}_1') = \langle \vec{r}_1 | \hat{\rho}_1 | \vec{r}_1' \rangle = N \sum_n \int \cdots \int \prod_{i=2}^{N} d\vec{r}_i \, \Psi_n^*(\vec{r}_1, \vec{r}_2, \ldots, \vec{r}_N) \, e^{-\beta \hat{H}} \, \Psi_n(\vec{r}_1', \vec{r}_2, \ldots, \vec{r}_N)$$

$$= \langle \hat{\Psi}^\dagger(\vec{r}_1) \, \hat{\Psi}(\vec{r}_1') \rangle = \frac{1}{V} \sum_{\vec{p}_1 \vec{p}_1'} e^{+i\hbar(\vec{p}_1 \vec{r}_1 - \vec{p}_1' \vec{r}_1')} \langle \hat{a}_p^\dagger \, \hat{a}_{p'} \rangle \tag{11.3}$$

where $\{\Psi_n\}$ is a set of complete orthonormal functions, and $\hat{\Psi}^\dagger(\vec{r})$ and $\hat{\Psi}(\vec{r})$ are quantized field operators.

If $\vec{P}$ is the total momentum operator given by

$$\hat{\vec{P}} = \sum_p \vec{p} \, \hat{a}_p^\dagger \, \hat{a}_p \tag{11.4}$$

Then,

$$\langle [\hat{\vec{P}}, \hat{a}_p^\dagger \, \hat{a}_{p'}] \rangle = Tr([\hat{w}, \hat{\vec{P}}] \, \hat{a}_p^\dagger \, \hat{a}_p) \tag{11.5}$$

If we expand the commutator of the right-hand member of Eq. (11.5), express the trace of the sum of operators as a sum of traces of operators and use the cyclic invariance property of the trace, we get

$$\langle [\hat{\vec{P}}, \hat{a}_p^\dagger \, \hat{a}_{p'}] \rangle = Tr[(\hat{w}, \hat{\vec{P}}) \, \hat{a}_p^\dagger \, \hat{a}_{p'}] \tag{11.6}$$

Since $\hat{w} = e^{-\beta \hat{H}} / Tr(e^{-\beta \hat{H}})$ and $[\hat{\vec{P}}, \hat{H}] = 0$, by assumption, we have

$$\langle [\hat{\vec{P}}, \hat{a}_p^\dagger \, \hat{a}_{p'}] \rangle = 0 \tag{11.7}$$

Using the commutation rules for bosons

$$\hat{a}_p \, \hat{a}_{p'}^\dagger - \hat{a}_{p'}^\dagger \, \hat{a}_p = \delta_{pp'}$$

$$\hat{a}_p \hat{a}_{p'} - \hat{a}_{p'} \, \hat{a}_p = 0$$

$$\hat{a}_p^\dagger \, \hat{a}_{p'}^\dagger - \hat{a}_{p'}^\dagger \hat{a}_p^\dagger = 0$$

we get

$$\langle [\hat{\vec{P}}_1, \hat{a}_p^\dagger \hat{a}_{p'}] \rangle = (\vec{p} - \vec{p}') \hat{a}_p^\dagger \hat{a}_{p'} \tag{11.8}$$

Hence from Eqs. (11.6) and (11.8), we get

$$\langle \hat{a}_p^\dagger \hat{a}_{p'} \rangle = 0 \quad \text{if } \vec{p} = \vec{p}' \tag{11.9}$$

We can thus write

$$\langle \hat{a}_p^\dagger \hat{a}_{p'} \rangle = \delta_{pp'} \langle n_p \rangle \tag{11.10}$$

where $\hat{n}_p = \hat{a}_p^\dagger \, \hat{a}_p$ is the occupation number operator whose eigenvalue is the occupation number n_p and the eigenvectors $|n_{p_1} n_{p_2} \ldots \rangle$. It follows from Eqs. (11.10) and (11.3) that

$$\rho_1(\vec{r}_1, \vec{r}_1') = \frac{1}{V} \sum_p e^{i\hbar \, \vec{p}(\vec{r}_1 - \vec{r}_1')} \langle \hat{n}_p \rangle \tag{11.11}$$

and

$$\frac{\langle n_0 \rangle}{V} = \underset{\substack{\vec{r} \to \infty \\ V \to \infty}}{\text{Lt}} \rho_1(\vec{r}) \tag{11.12}$$

where $\vec{r} = \vec{r}_1 - \vec{r}_1'$. At absolute zero,

$$\begin{aligned} w_n &= 1 \qquad \text{if } n = 0 \\ &= 0 \qquad \text{otherwise} \end{aligned} \tag{11.13}$$

and the ground state wave function Ψ_0 for helium is real, non-negative and always unique. We then have from Eq. (11.2)

$$\rho_1(\vec{r}) = N \int \cdots \int \prod_{i=2}^{N} d\vec{r}_i \, \Psi_0(\vec{r}, \vec{r}_2, \ldots, \vec{r}_N) \Psi_0(0, \vec{r}_2, \ldots, \vec{r}_n) \tag{11.14}$$

If we assume liquid He II at absolute zero as a Bose system of hard-spheres of diameter a and of the observed density (2.2×10^{22} atoms/cm^3) of liquid He^4, we can approximate Ψ_0 to the form

$$\Psi_0(\vec{r}, \vec{r}_2, \ldots, \vec{r}_N) = \frac{1}{\sqrt{Z_N}} \chi_N(\vec{r}, \vec{r}_2, \ldots, \vec{r}_N) \tag{11.15}$$

where

$$\begin{aligned} \chi_N &= 0 \qquad \text{if any } |\vec{r}_{ij}| \leq a \\ &= 1 \qquad \text{otherwise} \end{aligned} \tag{11.16}$$

where Z_N is a normalization constant, numerically equal to the configuration integral of a classical hard-sphere system. It is seen from the boundary conditions given by Eq. (11.16) that

$$\chi_N(\vec{r}, \vec{r}_2, \ldots, \vec{r}_N) \, \chi_N(0, \vec{r}_2, \ldots, \vec{r}_N) = \chi_{N+1}(\vec{r}, 0, \vec{r}_2, \ldots, \vec{r}_N) \tag{11.17}$$

Substitution of Eqs. (11.17) and (11.15) in Eq. (11.14) gives

$$\rho_1(\vec{r}) = \frac{N}{Z_N} \int \cdots \int \prod_{i=2}^{N} d\vec{r}_i \, \chi_{N+1}(\vec{r}, 0, \vec{r}_2, \ldots, \vec{r}_N) \tag{11.18}$$

The canonical 2-particle density distribution function $\rho_2(\vec{r})$ for a hard-sphere uniform fluid of $(N + 1)$ particles, with the aid of Eq. (3.55), is given by

$$\rho_2(\vec{r}) = \frac{(N+1)N}{Z_{N+1}} \int \cdots \int \prod_{i=2}^{N} d\vec{r}_i \, \chi_{N+1}(\vec{r}, 0, \vec{r}_2, \ldots, \vec{r}_N) \tag{11.19}$$

We know that

$$\underset{r \to \infty}{\text{Lt}} \rho_2(\vec{r}) \to \left(\frac{N}{V}\right)^2 \tag{11.20}$$

Comparing Eqs. (11.18) and (11.19) and using Eq. (4.12), we get

$$\rho_1(\vec{r}) = \frac{\rho_2(\vec{r})}{z} \tag{11.21}$$

where z is fugacity and is equal in terms of configuration integrals to

$$z = \frac{Z_N/N!}{Z_{N+1}(N+1)!}$$

If follows from Eqs. (11.20), (11.12) and (11.21) that

$$\frac{\langle n_0 \rangle}{N} = \frac{N}{Vz} = \frac{\rho}{z} \tag{11.22}$$

This shows that Bose–Einstein condensation exists at absolute zero if z is finite.

With the aid of Eq. (8.42) we have

$$\frac{\langle n_0 \rangle}{N} = \exp\left(\sum_{k \geq 1} \beta_k \rho^k \right) \tag{11.23}$$

The right-hand member of Eq. (11.23) can be calculated after evaluating the irreducible cluster integrals. Exact numerical calculation of the first few virial coefficients of a hard-sphere classical system is feasible (Hirshfelder, Curtiss and Bird, *Molecular Theory of Gases and Liquids*, John Wiley & Sons, N.Y., London 1954, Chapter III page 157). We quote some of the results:

$$B_2 = \frac{2\pi a^3}{3}, \frac{B_3}{B_2^2} = \frac{5}{8}, \frac{B_4}{B_2^3} = 0.28695, \frac{B_5}{B_2^4} = 0.1103, \ldots$$

where a is the hard sphere diameter. From these, the irreducible cluster integrals β_k can be calculated. Taking a = 2.556 Å, $\rho = 2.160 \times 10^{22}$ atom/cm^3 and evaluating β_k up to $k = 4$, we get

$$\frac{\langle n_0 \rangle}{N} \simeq 0.08 \tag{11.24}$$

THE TWO-FLUID MODEL

Tisza (1938) employed the qualitative properties of a degenerating Bose–Einstein gas to develop a consistent macroscopic theory called the *two-fluid theory*, which provides a common theoretical basis for the most striking facts about liquid He II. The two-fluid theory of Tisza and London is based on the following assumptions:

1. Liquid helium consists of two mutually interpenetrating fluids, the *superfluid* of mass density ρ_s and the *normal fluid* of mass density ρ_n. Thus, at any point in space, the mass density and the mass current density can be written in the form

$$\rho = \rho_s + \rho_n \tag{11.25}$$

$$\vec{J} = \rho\vec{v} = \rho_s\vec{v}_s + \rho_n\vec{v}_n \tag{11.26}$$

2. The superfluid corresponds to the condensed phase of the Bose–Einstein fluid. Thus, ρ_s increases from 0 to ρ as the temperature decreases from 2.18 K (the λ-point) to 0 K.

The superfluid as a whole is virtually one quantum state and does not contribute to the entropy

$$S_s = 0 \tag{11.27}$$

Further, because of the superfluid being represented by one quantum state, there is no collision mechanism of the usual kind within the superfluid and the viscosity coefficient of the superfluid is zero.

$$\eta_s = 0 \tag{11.28}$$

3. The normal fluid is the carrier of the whole thermal excitation. Thus, the entropy of the liquid He II is entirely attributed to the normal liquid.

$$\rho S = \rho_n S_n \tag{11.29}$$

where S and S_n refer to entropies per gram. The normal fluid has a viscosity, which goes over continuously to the value of liquid He I at the λ-point

$$\underset{T \to T_\lambda}{\text{Lt}} \eta_n(T) = \eta_I(T_\lambda) \tag{11.30}$$

The two-fluid model explained many of the observations. Keesom and MacWood, with the rotating disc method, found that the viscosity of liquid helium below T_λ decreases with decreasing temperature quite considerably and continuously and at 1.5 K has a value one-tenth of the value just above T_λ. In contrast to it, Kapitza and independently Allen and Misener using capillary flow method found that the viscosity was immeasurably small below the λ point. The two-fluid method resolves this contradiction in a very simple way. In the capillary flow method, the flow is mainly realized by the superfluid, the normal fluid being held back by its viscosity. On the other hand, the rotating disc interacts with the normal fluid only and is not affected by the superfluid. Thus, the two experiments deal with two different fluids and there is no contradiction. The existence of the two fluids and measurement of the ratio ρ_n/ρ was made by the experiment of Andronikashvilli (1946). He used a pile of 100 aluminium sheets (0.0013 cm thick) at a distance of 0.021 cm from each other, packed in a cylinder of diameter 3.5 cm and length 2.35 cm suspended by a thin phosphor bronze wire (giving rise to the restoring force). When discs oscillate, only the normal fluid moves between them and contributes to the moment of inertia. For each temperature,

$$\frac{\text{The experimental moment of inertia}}{\text{The geometrical moment of inertia}} = \frac{\rho_n}{\rho}$$

It was observed that ρ_n/ρ is unity at T_λ and ρ_n/ρ virtually seems to disappear below 1 K. It was observed that between 1.3 K to 2.18 K,

$$\frac{\rho_n}{\rho} = \left(\frac{T}{T_\lambda}\right)^{5.6}$$

If two containers A and B containing liquid helium are connected under a pressure differential by a capillary tube so fine that the normal fluid is immobilized and suppose some superfluid from A flows to B, the vessels being perfectly isolated from each other and from the outside, the entropy per unit mass of container A will increase and that of B will decrease, as superfluid has no entropy. This means that the temperature will increase in A and decrease in B. This is known as *mechanocaloric effect.* The effect was not known then, but was inferred by Tisza in 1938. The effect was discovered by Daunt and Mendelssohn in 1939 and was measured by Kapitza in 1941, who used a narrow slit (width 10^{-5} cm) formed by two optically polished quartz discs. The inverse effect, known as *fountain effect* (thermo-mechanical effect), however, was discovered by Allen and Jones in 1938. Their apparatus consisted of a U-tube packed with emery powder, having one arm lengthened in a fine capillary tube, the U-tube being immersed in liquid He II. When allowed light falls on the portion containing emery powder creating a temperature gradient, fountain flow is observed at the capillary end. This happens because the temperature difference produces a pressure difference, the higher temperature being associated with higher pressure.

The most notable prediction from the two-fluid theory was that of the second sound. The familiar (first) sound is a pressure wave in which the two fluids move in phase (oscillations of ρ_n and ρ_s) with each other. Another type of wave in which the two fluids vibrate with a phase difference of π, a compression wave of entropy accompanied by the periodic fluctuations of the temperature, the mass density remaining almost constant, is possible. This mode of oscillation is excited by local heating of liquid He II. The temperature gradient produced will not propagate by diffusion (as in heat conduction), but will propagate like a wave with a characteristic velocity. The phenomenon is called the *second sound.*

The wave equation and consequently the velocity of the thermal waves (the second sound) can be derived in an elementary way (though not rigorously) as below.

As there is no net transport of mass across any plane in the liquid, Eq. (11.26) reduces to

$$\rho_s\vec{v}_s + \rho_n\vec{v}_n = 0 \tag{11.31}$$

The superfluid has zero entropy, hence the equation of conservation of entropy can be written as

$$\frac{\partial(\rho S)}{\partial t} + \operatorname{div}(\rho S\vec{v}_n) = 0 \tag{11.32}$$

Consider a layer of liquid helium of thickness dx in which the temperature varies by dT, the transfer of heat per unit time is

$$Q = \rho ST\vec{v}_n \tag{11.33}$$

If work W is done reversibly by transferring the heat Q from T to $T - dT$, the second law of thermodynamics requires

$$W = Q\frac{dT}{T} = \rho S\vec{v}_n\, dT \tag{11.34}$$

The kinetic energy E of 'internal convection' (there is no mass transfer) is

$$E = \frac{1}{2}(\rho_n\vec{v}_n^2 + \rho_s\vec{v}_s^2) = \frac{1}{2}\rho\frac{\rho_n}{\rho_s}\vec{v}_n^2 \tag{11.35}$$

(where we use Eq. (11.31) and $\rho = \rho_s + \rho_n$).

According to Tisza, the work W brings about the change of the kinetic energy of internal convection per unit area of the layer of thickness dx, i.e.

$$dE\,dx = -W\,dt = -\rho S\vec{v}_n\,dT\,dt$$

or

$$\frac{\partial E}{\partial t} = -\rho S\vec{v}_n \text{ grad } T \tag{11.36}$$

Substitution of E from Eq. (11.35) in Eq. (11.36) yields

$$\rho\frac{\rho_n}{\rho_s}\vec{v}_n\dot{\vec{v}}_n = -\rho S\vec{v}_n \text{grad } T \tag{11.37}$$

We have

$$C_v \text{ grad } T = T \text{ grad } S \tag{11.38}$$

From Eqs. (11.37) and (11.38), we get

$$\dot{\vec{v}}_n + \frac{\rho_s}{\rho_n}\frac{T}{C_v} S \text{ grad } S = 0 \tag{11.39}$$

Eliminating $\vec{v}_n$ from Eqs. (11.39) and (11.32), considering only terms linear in $\vec{v}_n$, grad T, grad S, S, etc. as small of first order and neglecting products of any of these two quantities, we get

$$\frac{\partial^2 S}{\partial t^2} - \frac{\rho_s}{\rho_n} S^2 \frac{T}{C_v} \text{ div grad } S = 0 \tag{11.40}$$

This is a wave equation for entropy. The same equation also holds for temperature. The propagation velocity for these thermal waves (the velocity of the second sound) is

$$v_{\text{II}}^2 = \frac{\rho_s}{\rho_n} S^2 \frac{T}{C_v} \tag{11.41}$$

Peshkov in 1944 succeeded in experimentally producing thermal waves in liquid He II and measuring its velocity as function of temperature in a wide temperature range. He found v_{II} to have a maximum of 20.36 m/s at 1.65 K (very close to maximum of the theoretical curve), a minimum of about 18.4 m/s at 1.1 K and rise again if the temperature is lowered below 1.1 K. Atkins and Osborne (1950) found a rapid increase of v_{II} tending towards a finite limit of about 150 m/s.

The two fluid theory is simple and it works in explaining the striking facts of liquid He II. However, a complete understanding of liquid helium is not possible because the two fluid theory lacks a complete hydrodynamic description and the nature of the two fluids in molecular terms. The prediction of viscosity and the velocity of second sound as $T \to 0$ K are not satisfactory. These deficiencies are partly removed by the theories of Landau and Feynman.

LANDAU THEORY OF LIQUID He II

The theory of Landau (Journal of Physics, U.S.S.R., **5**, 71, 1941) tries to lay a molecular foundation for the two-fluid theory.

Landau rejected the approach from the side of gas statistics (which London and Tisza followed). He approached the problem from the solid side (in which Bose–Einstein condensation has no relevance). At zero temperature, the atoms of the solid body are in the ground state (the lowest energy state). At higher temperatures, the atoms undergo oscillations around their equilibrium positions, which are quantized as phonons. These phonons correspond to sound waves in the same way as photons correspond to light waves. These elementary particles behave like quasi-particles of definite energy and momentum. Landau assumed that the excited state of liquid He II can be represented by two kinds of elementary excitations, the *phonons* and the *rotons*. The phonons are excitations of the longitudinal sound waves and represent the potential motions of the liquid, characterized by curl $\vec{v} = 0$. The rotons are the elementary excitations of the vortex spectrum, characterized by curl $\vec{v} \neq 0$.

The energy of the phonons is a linear function of the momentum

$$\in \; = v_l p \qquad (11.42)$$

To explain the experimental values of the thermodynamic functions, Landau proposed the energy spectrum of the type shown below (Fig. 11.4).

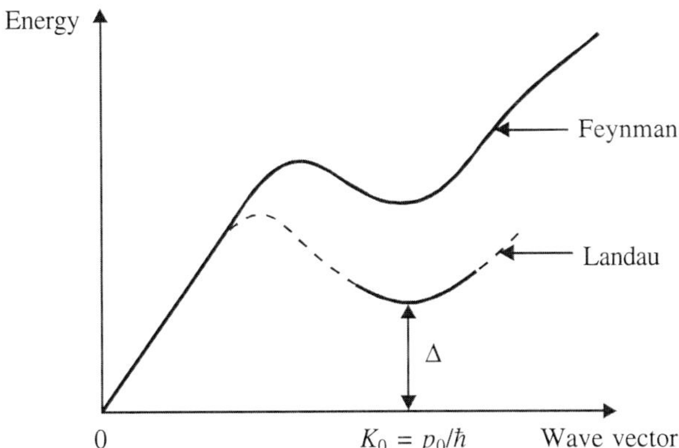

Figure 11.4 Dispersion curve for liquid He^4.

The energy of the rotons are represented near the minimum by

$$\in = \Delta + \frac{(p - p_0)^2}{2\mu} \qquad (11.43)$$

where p_0 is the value of the momentum at which energy has a minimum equal to Δ and μ is the effective mass of the roton.

From the experiments of neutron scattering, the whole energy spectrum of the excitation is determined, giving the following values of the parameters:

$$v_I = 226 \text{ m/s}, \; \frac{\Delta}{k} = 8.6 \text{ K}, \; \frac{p_0}{\hbar} = 1.91 \text{ Å}^{-1}, \quad \mu = 0.16 \; m_{He}$$

Both phonons and rotons have integral spins and obey Bose statistics. The roton energy has a large quantity Δ (as compared to kT), and hence Bose distribution can be replaced by Boltzmann distribution in the case of rotons.

It can be shown that the property of superfluidity follows from the motion of elementary excitations. Consider the liquid helium at zero temperature, i.e., in the ground state and the liquid flow through a capillary with velocity $\vec{v}$. If this flow were accompanied by friction, a part of the kinetic energy will be dissipated and converted into thermal energy, thereby heating up the liquid and making transition to excited states. If $\in(p)$ is the energy of an excitation with momentum $\vec{p}$ in the frame moving with the liquid, in the fixed frame, the energy changes by an amount

$$\in(p) + \vec{p}\cdot\vec{v}$$

The transition to the excited state is possible if

$$\in(p) + \vec{p}\cdot\vec{v} < 0 \tag{11.44}$$

The favourable situation being one in which the momentum of the created excitation is directed opposite to the velocity. We then have

$$v > \frac{\in(p)}{p} \tag{11.45}$$

Obviously, an excitation can appear if Eq. (11.45) is fulfilled, at least at that point in the spectrum where $\in(p)/p$ has a minimum value. Hence the necessary condition for the creation of an excitation is

$$v > \min \frac{\in(p)}{p} \tag{11.46}$$

or

$$v_{\text{critical}} = \min \frac{\in(p)}{p} \neq 0 \tag{11.47}$$

For the values of v, which are less than v_{critical}, no excitation is created, i.e. the liquid will flow without dissipation of energy, i.e. without friction, i.e. it will be a superfluid. The criterion given by Eq. (11.46) is true at finite temperatures also. The presence of excitations in the liquid at finite temperature introduces special features into the flow of the liquid through the capillary. The excitations are reflected against the walls and thereby transfer a part of their momentum to the wall. Because of this, that part of the liquid which is carried along by the motion of the excitations, behaves like a normal viscous liquid and is slowed down by friction with the walls. Thus at $T = 0$ K, the whole liquid flows frictionless (superfluid) whereas at $T \neq 0$ K, only a part does, i.e. we have two components, the superfluid and the normal fluid, which satisfy the same equations as in Tisza theory that

$$\rho = \rho_n + \rho_s \tag{11.48}$$

and

$$\vec{J} = \rho\vec{v} = \rho_s\vec{v}_s + \rho_n\vec{v}_n \tag{11.49}$$

The ratio ρ_n/ρ is unity at $T = T_\lambda$ and is zero at $T = 0$ K.

Both phonons and rotons contribute to the thermodynamic properties of liquid He II. At temperatures below 1 K, phonons play a dominant role as few rotons exist. Above 1 K temperature, the roton contribution becomes important. We calculate their contribution to the thermodynamic functions, assuming them to be ideal gases as their number is not large (smaller densities).

As the phonon gas obeys Bose statistics, the energy per unit volume of the phonon gas is given by

$$E_{\text{ph}} = \frac{4\pi}{h^3}\int_0^\infty \frac{\epsilon p^2 dp}{e^{v_{\text{I}}p/kT}-1} = \frac{4\pi v_{\text{I}}}{h^3}\int_0^\infty \frac{p^3 dp}{e^{v_{\text{I}}p/kT}-1}$$

$$= \frac{4\pi}{h^3 v_{\text{I}}^3}(kT)^4 \int_0^\infty \frac{x^3 dx}{e^x - 1} \tag{11.50}$$

where $x = v_{\text{I}}p/kT$.

Using integral in Eq. (7.9), we have

$$E_{\text{ph}} = \frac{4}{15}\frac{\pi^5}{h^3 v_{\text{I}}^3}(kT)^4 = \frac{bT^4}{4} \tag{11.51}$$

where

$$b = \frac{16\pi^5 k^4}{15 h^3 v_{\text{I}}^3} \tag{11.52}$$

The specific heat per unit volume

$$C_{\text{ph}} = \left(\frac{\partial E}{\partial T}\right)_V = bT^3 = \frac{16\pi^5 k^4}{15 h^3 v_{\text{I}}^3} T^3 \tag{11.54}$$

and the entropy per unit volume

$$S_{\text{ph}} = \int \frac{C_{\text{ph}}}{T} dT = \frac{b}{3}T^3 = \frac{16\pi^5 k^4}{45 h^3 v_{\text{I}}^3} T^3 \tag{11.55}$$

The roton gas obeys Boltzmann statistics; the free energy is given by

$$A = -kT \ln Q_N = -kT \ln \frac{q^N}{N!} = -NkT \ln q + NkT \ln N - NkT \tag{11.56}$$

where q is the single particle partition function, given by

$$q = \frac{4\pi V}{h^3}\int_0^\infty \exp\left(-\frac{\epsilon}{kT}\right) p^2 dp = \frac{4\pi V}{h^3}\int_0^\infty \exp\left(-\frac{\Delta + (p-p_0)^2/2\mu}{kT}\right) p^2 dp$$

We substitute $p - p_0 = t$, $dp = dt$ and $p^2 = (p_0 + t)^2 \simeq p_0^2$ since $p_0^2/2\mu kT >> 1$, to obtain

$$q = \frac{4\pi V}{h^3}\exp\left(-\frac{\Delta}{kT}\right) p_0^2 \int_{-\infty}^\infty \exp\left(-\frac{t^2}{2\mu kT}\right) dt = \frac{4\pi V}{h^3} p_0^2 \exp\left(-\frac{\Delta}{kT}\right)\sqrt{2\pi\mu kT} \tag{11.57}$$

In the above, the lower limit has been extended from $-p_0$ to $-\infty$.

The number of rotons changes with temperature; it has to be determined from the condition that A is minimum, i.e.

$$\left(\frac{\partial A}{\partial N}\right)_V = 0 = -kT \ln q + kT \ln N \tag{11.58}$$

which gives the number of rotons per unit volume

$$N_{\text{rot}} = \frac{q}{V} = \frac{4\pi}{h^3} p_0^2 \sqrt{2\pi\mu kT}\, e^{-\Delta/kT} \tag{11.59}$$

The free energy per unit volume is

$$A_{\text{rot}} = -kT\, N_{\text{rot}} = -\frac{4\pi}{h^3} p_0^2 \sqrt{2\pi\mu kT}\, kT\, e^{-\Delta/kT} \tag{11.60}$$

The entropy per unit volume

$$S_{\text{rot}} = -\left(\frac{\partial A_{\text{rot}}}{\partial T}\right)_V = kN_{\text{rot}}\left(\frac{3}{2} + \frac{\Delta}{kT}\right) \tag{11.61}$$

and the specific heat per unit volume

$$C_{\text{rot}} = T\left(\frac{\partial S}{\partial T}\right)_V = kN_{\text{rot}}\left[\left(\frac{\Delta}{kT}\right)^2 + \frac{\Delta}{kT} + \frac{3}{4}\right] \tag{11.62}$$

Adding up the contributions due to phonon gas and roton gas, we find the expressions for entropy and specific heat of the liquid He II

$$S = S_{\text{ph}} + S_{\text{rot}} = \frac{16\pi^5 k^4}{45 h^3 v_{\text{I}}^3} T^3 + kN_{\text{rot}}\left(\frac{\Delta}{kT} + \frac{3}{2}\right) \tag{11.63}$$

$$C = C_{\text{ph}} + C_{\text{rot}} = \frac{16\pi^5 k^4}{15 h^3 v_{\text{I}}^3} T^3 + k\, N_{\text{rot}}\left[\left(\frac{\Delta}{kT}\right)^2 + \frac{\Delta}{kT} + \frac{3}{4}\right] \tag{11.64}$$

For liquid He II in the capillary tube along z-direction, the energy of the phonon in the fixed frame is $\in - v_z p \cos\theta$, where θ is the angle that $\vec{p}$ makes with z-direction. Then the average momentum per unit volume

$$\langle p \rangle = \frac{1}{V}\int p\cos\theta\, dN_{\text{ph}} = \frac{1}{h^3}\int_0^\infty\int_0^\pi\int_0^{2\pi} \frac{p\cos\theta}{e^{(\in - v_z p\cos\theta)/kT} - 1} p^2 dp \sin\theta\, d\theta\, d\phi$$

Since v_z is small, we use a Taylor expansion and retaining only the linear term, we get

$$\frac{1}{e^{(\in - v_z p\cos\theta)/kT} - 1} \simeq \frac{1}{e^{\in/kT} - 1} + \frac{v_z (p\cos\theta/kT) e^{\in/kT}}{(e^{\in/kT} - 1)^2}$$

When we substitute the above expansion in <p>, the first integral vanishes because $\int_0^{\pi} \sin\theta \cos\theta d\theta = 0$, in the second integral we use $\int_0^{\pi} \sin\theta \cos^2\theta d\theta = \frac{2}{3}$ and $\in = p(v_{\rm I} + v_z \cos\theta) \simeq pv_{\rm I}$ since $v_z \ll v_{\rm I}$. This results in

$$\langle p \rangle = \frac{4\pi}{3h^3} \frac{v_z}{kT} \int_0^{\infty} \frac{e^{pv_{\rm I}/kT}}{(e^{pv_{\rm I}/kT} - 1)^2} p^4 dp$$

We carry out the integration by parts; the first part vanishes for both the limits, giving us

$$\langle p \rangle = \frac{4}{3} \frac{E_{\rm ph}}{v_{\rm I}^2} v_z$$

We change from per unit volume to per unit mass by putting $v = 1/r$ and write $\langle p \rangle = (\rho_{\rm ph}/\rho)v_z$ to get

$$\rho_{\rm ph} = \rho \frac{4}{3} \frac{E_{\rm ph}}{v_{\rm I}^2} = \frac{\rho b T^4}{3v_{\rm I}^2} \tag{11.65}$$

For $T < 1$ K, the normal fluid is entirely made of the phonon gas, hence writing $\rho_s = \rho - \rho_n = \rho - \rho_{\rm ph}$ and substituting this in Eq. (11.41) and for S and C_v from Eqs. (11.55) and (11.54), we get

$$v_{II} = \frac{1}{3}(3v_I^2 - bT^4)^{1/2} \tag{11.66}$$

As $T \to 0$, $v_{\rm II} \simeq v_{\rm I}/\sqrt{3} = 137$ m/s

The above is almost in agreement with experiments of Atkins and Osborne and corrects the anomaly of Tisza theory that $v_{\rm II} \to 0$ as $T \to 0$ K.

We proceed now to the quantization of the motion of the liquid. A classical liquid is described by the density ρ and the current vector $\vec{J}$ defined as

$$\rho = \sum_{\alpha} m_{\alpha} \delta(\vec{r} - \vec{r}_{\alpha}) \tag{11.67}$$

$$\vec{J} = \sum_{\alpha} \vec{p}_{\alpha} \delta(\vec{r} - \vec{r}_{\alpha}) = \sum_{\alpha} m_{\alpha} \vec{v}_{\alpha} \delta(\vec{r} - \vec{r}_{\alpha}) \tag{11.68}$$

where m_{α}, $\vec{v}_{\alpha}$, $\vec{p}_{\alpha}$ are the mass, velocity and momentum of the αth-particle and the sum is taken over all the particles of the system.

When we go over to quantum theory, ρ and $\vec{J}$ must be regarded as operators whose form must be determined. The operator ρ for one particle is defined in such a manner that its expectation value $\int \psi^*(\vec{r}_{\alpha}) \rho \psi(\vec{r}_{\alpha}) dV$ is equal to $m_{\alpha} \psi^*(\vec{r})\, \psi(\vec{r})$. The operator satisfies this condition. Hence for many particles operator, ρ has the form

$$\rho = \sum_{\alpha} m_{\alpha} \delta(\vec{r} - \vec{r}_{\alpha}) \tag{11.69}$$

In quantum mechanics, the current vector is given by

$$-\frac{1}{2} i\hbar[\psi^*(\vec{r})\nabla\psi(\vec{r})-\psi(\vec{r})\nabla\psi^*(\vec{r})] \tag{11.70}$$

The corresponding symmetrized quantum operator for the current density given by Eq. (11.68) is

$$\vec{J}=\frac{1}{2}\{\vec{p}\,\delta(\vec{r}-\vec{r}_\alpha)+\delta(\vec{r}-\vec{r}_\alpha)\vec{p}\}$$

where $\vec{p}\to -i\hbar\nabla$. We can easily see that the expectation value of the operator defined above is equal to Eq. (11.70). Hence many particles current operator has the form

$$\vec{J}=\frac{1}{2}\sum_\alpha\{\vec{p}_\alpha\delta(\vec{r}-\vec{r}_\alpha)+\delta(\vec{r}-\vec{r}_\alpha)\vec{p}_\alpha\} \tag{11.71}$$

In hydrodynamics, along with current, one uses the velocity $\vec{v}$. The corresponding operator is

$$\vec{v}=\frac{1}{2}\left(\frac{1}{\rho}\vec{J}+\vec{J}\frac{1}{\rho}\right) \tag{11.72}$$

The commutation relations between these operators can be obtained directly by calculating the commutators. The results are

$$\vec{J}_1\rho_2-\rho_2\vec{J}_1=-i\hbar\rho_1\nabla\delta(\vec{r}_1-\vec{r}_2) \tag{11.73}$$

$$\rho_1\rho_2-\rho_2\rho_1=0 \tag{11.74}$$

$$\vec{v}_1\rho_2-\rho_2\vec{v}_1=-i\hbar\nabla\delta(\vec{r}_1-\vec{r}_2) \tag{11.75}$$

where the subscripts 1 and 2 denote the values of the operators at points $\vec{r}_1$ and $\vec{r}_2$.

The commutation rules for the components of $\vec{v}$ are

$$v_{1i}v_{2k}-v_{2k}v_{1i}=-i\hbar\,\delta(\vec{r}_1-\vec{r}_2)\frac{1}{\rho_1}(\text{curl}\,\vec{v})_{ik} \tag{11.76}$$

Also, we have

$$\text{curl}\,\vec{v}_1\rho_2-\rho_2\,\text{curl}\,\vec{v}_1=0 \tag{11.77}$$

By applying the above formula to the macroscopic movement of the liquid, we get the hydrodynamic equations written in the operator form. The energy per unit volume macroscopically in a classical liquid is

$$\frac{\rho v^2}{2}+E(\rho) \tag{11.78}$$

where $E(\rho)$ is the internal energy expressed as a function of ρ only. The corresponding quantum operator has the form

$$\frac{\vec{v}\cdot\rho\vec{v}}{2}+E(\rho) \tag{11.79}$$

The Hamiltonian H is an integral over the volume

$$H = \int \left\{ \frac{\vec{v} \cdot \rho \vec{v}}{2} + E(\rho) \right\} dV \tag{11.80}$$

The derivative of ρ with respect to time is given by

$$\dot{\rho} = \frac{i}{\hbar}(H\rho - \rho H) \tag{11.81}$$

Inserting the value of H from Eq. (11.80) into Eq. (11.81), we get the continuity equation in operator form

$$\dot{\rho} + \text{div}\, \frac{\rho \vec{v} + \vec{v} p}{2} = 0 \tag{11.82}$$

In an analogous manner, we can obtain

$$\dot{\vec{v}} = \frac{i}{\hbar}(H\vec{v} - \vec{v}H) \tag{11.83}$$

from which the operator equation of motion follows:

$$\dot{v}_i + \left(v_k \frac{\partial v_i}{\partial x_k} + \frac{\partial v_i}{\partial x_k} v_k \right) = -\frac{1}{\rho} \nabla P \tag{11.84}$$

where $P = \rho^2(\partial/\partial\rho)[(E/\rho)]$, the classical expression for pressure.

The choice of the values of Δ, p_0 and μ_r gives a good fit for the values of C_v, S and ρ_n/ρ with experimental results. But Landau theory does not explain the existence of T_λ satisfactorily. Also, Landau theory gives only the linear part and the parabola part (shown by solid line in Fig. 11.4) of the energy spectrum and not the whole spectrum.

FEYNMAN THEORY OF LIQUID He II

Feynman theory (*Phys. Rev.*, **94**, 262, 1954) throws more light on the nature of the elementary excitations. Feynman assumes the wave function of the form

$$\psi = \sum_{i=1}^{N} f(\vec{r}_i)\psi_0 \tag{11.85}$$

where ψ_0 is the wave function of the ground state and $f(r)$ is a function to be determined by minimizing the energy of the state. The Hamiltonian of the liquid helium is

$$\hat{H} = -\frac{\hbar^2}{2m} \sum_{i=1}^{N} \nabla_i^2 + \sum_{1 \le i < j \le N} u(r_{ij}) \tag{11.86}$$

where $u(r_{ij})$ is the interaction potential between the particles at $\vec{r}_i$ and $\vec{r}_j$ (the total interaction potential assumed to be expressible as a sum of pair interaction potentials) and m is the mass of He^4 atom. We introduce the function F defined by

$$F = \sum_{i=1}^{N} f(\vec{r}_i) \tag{11.87}$$

Then,

$$(\hat{H}-E_0)\psi = \hat{H}F\psi_0 - E_0F\psi_0$$
$$= -\frac{\hbar^2}{2m}\sum_{i=1}^{N}[(\nabla_i^2 F)\psi_0 + 2(\nabla_i F)\cdot(\nabla_i\psi_0)] \tag{11.88}$$

where we have used $\hat{H}\psi_0 = E_0\psi_0$.

If we write

$$\psi_0^2(\vec{r}_1,\ldots,\vec{r}_N) = \rho_N \tag{11.89}$$

Then

$$(\hat{H}-E_0)\psi = \frac{1}{\psi_0}\left[-\frac{\hbar^2}{2m}\sum_{i=1}^{N}\nabla_i(\rho_N\nabla_i F)\right] \tag{11.90}$$

The expectation value of $(\hat{H}-E_0)$ is the energy of the excitation and will come from minimizing the integral

$$\varepsilon = \int \psi^*(\hat{H}-E_0)\psi\, d\vec{r}_1\ldots d\vec{r}_N$$
$$= -\frac{\hbar^2}{2m}\int d\vec{r}_1\ldots d\vec{r}_N F^*\sum_{i=1}^{N}\nabla_i(\rho_N\nabla_i F)$$
$$= \frac{\hbar^2}{2m}\sum_{i=1}^{N}\int d\vec{r}_1\ldots d\vec{r}_N\, \nabla_i F^*\nabla_i F\rho_N$$

The integral of ρ_N over all atomic coordinates except $\vec{r}_i$ gives a result involving only $\rho_1(\vec{r}_i)=\rho_0$ (the probability of finding an atom at $\vec{r}_i$ in the ground state of the liquid). This gives

$$\varepsilon = \frac{\hbar^2}{2m}\rho_0\int d\vec{r}\,\nabla f^*(r)\nabla f(r) \tag{11.91}$$

where

$$\rho_0 = \int d\vec{r}_2\ldots d\vec{r}_N\rho_N \tag{11.92}$$

The normalization integral

$$I = \int d\vec{r}_1\ldots d\vec{r}_N\,\psi^*\psi = \sum_{i=1}^{N}\sum_{j=1}^{N}\int d\vec{r}_1\ldots d\vec{r}_N f^*(\vec{r}_i)f(\vec{r}_j)\rho_N$$

Fixing two values r_1 and r_2 and performing integration over all other coordinates:

$$I = \int f^*(\vec{r}_1)f(\vec{r}_2)\,\rho_2(\vec{r}_1,\vec{r}_2)\,d\vec{r}_1\,d\vec{r}_2$$

where ρ_2 is the probability of finding an atom at $\vec{r}_1$ per cm^3 and another at $\vec{r}_2$ per cm^3. We can write $\rho_2 = \rho_0 h(\vec{r}_1-\vec{r}_2)$, where $h(\vec{r}_1-\vec{r}_2)$ is the probability of finding an atom at $\vec{r}_2$ per unit volume if one is known to be at $\vec{r}_1$. This gives

$$I = \rho_0\int d\vec{r}_1\,d\vec{r}_2\,f^*(\vec{r}_1)f(\vec{r}_2)h(\vec{r}_1-\vec{r}_2) \tag{11.93}$$

The best choice of f is that which minimizes the ratio ε/I. The variation with respect to f^* gives the equation

$$E\int d\vec{r}_2 h(\vec{r}_1-\vec{r}_2)f(\vec{r}_2)=-\frac{\hbar^2}{2m}\nabla^2 f(\vec{r}_1) \tag{11.94}$$

where E is ε/I. This has the solution

$$f(\vec{r})=\text{constant } e^{i\vec{k}\cdot\vec{r}} \tag{11.95}$$

with the energy value

$$E(\vec{k})=\frac{\hbar^2k^2}{2mS(\vec{k})} \tag{11.96}$$

where $S(\vec{k})$ is the Fourier transform of the function $h(\vec{r})$

$$S(\vec{k})=\int d\vec{r}\,h(\vec{r})\,e^{i\vec{k}\cdot\vec{r}} \tag{11.97}$$

It is a function only of k; the magnitude of $S(\vec{k})$ is just the liquid structure factor, which is determined by the scattering of neutrons by the liquid at absolute zero. The experimental value of $S(k)$ will give $E(k)$. This leads to the graph shown in Fig. 11.4. It differs considerably from the Landau curve but the roton part comes out qualitatively. The quantitative agreement is poor but the notable feature of Feynman theory is that there are no ad hoc adjustable parameters. A better trial wave function has been devised by Feynman and Cohen (*Phys. Rev.*, **102**, 1189, 1956), which gives more quantitatively satisfactory results.

GROUND STATE ENERGY OF LIQUID He^4

The theories of Landau and Feynman establish an atomic basis on which the low temperature behaviour of liquid He^4 can be interpreted, but they are not rigorous. The main problem is the actual interparticle interaction potential whose exact form is not known, though different forms have been proposed for it. Once the energy eigenvalues are known, it can be used to evaluate the partition function and consequently discuss its macroscopic equilibrium and transport properties. The first attempts were made with a simpler model of the hard-sphere interaction. Bogoliubov (*J. Phys.* U.S.S.R., **11**, 23, 1947) obtained the excitation spectrum in the case of weak interactions, using perturbation technique. The Hamiltonian of the system in the language of quantized fields can be written as

$$\hat{H}=\sum\frac{\hat{p}^2}{2m}\hat{a}_p^\dagger\hat{a}_p+\frac{1}{2V}\sum V_{\hat{p}_1,\hat{p}_2;\hat{p}_3\hat{p}_4}\hat{a}_{p_1}^\dagger\hat{a}_{p_2}^\dagger\hat{a}_{p_3}\hat{a}_{p_4} \tag{11.98}$$

where $\hat{p}$ is momentum operator, $V_{\hat{p}_1,\hat{p}_2;\hat{p}_3\hat{p}_4}$ is the interaction matrix elements and the field operators of the boson field satisfy the commutation relations

$$[\hat{a}_p,\hat{a}_{p'}^\dagger]=\delta_{pp'}$$

$$[\hat{a}_p,\hat{a}_{p'}]=0=[\hat{a}_p^\dagger,\hat{a}_{p'}^\dagger] \tag{11.99}$$

Let us consider the system near absolute zero, almost all particles are in the condensate, i.e. $\hat{p} = 0$. For low lying excited states, i.e. small p, the interaction matrix element $V_{\hat{p}_1,\hat{p}_2;\hat{p}_3\hat{p}_4}$ can be replaced by a constant, say α. Also, the total number of particles N is equal to

$$N = N_0 + \sum_p n_p \tag{11.100}$$

with $\sum_p n_p \ll N_0$. The number of condensed particles N_0 being large, we can replace (classical treatment) $\hat{a}_0^\dagger$ and $\hat{a}_0$ by $N_0^{1/2}$.

The largest contribution to the interaction energy comes from the interaction of condensed particles amongst themselves and the interaction of the excited particles with the condensed particles. We can write these terms in the form

$$\frac{N_0^2}{2V}\alpha + \frac{N_0}{V}\alpha \sum (\hat{a}_p\hat{a}_{-p} + \hat{a}_p^\dagger\hat{a}_{-p}^\dagger + 2\hat{a}_p^\dagger\hat{a}_p + 2\hat{a}_{-p}^\dagger\hat{a}_{-p}) \tag{11.101}$$

Using Eq. (11.100) with $n_p = \hat{a}_p^\dagger\hat{a}_p$, Eq. (11.101) becomes

$$\frac{N^2}{2V}\alpha + \frac{N}{V}\alpha \sum (\hat{a}_p\hat{a}_{-p} + \hat{a}_p^\dagger\hat{a}_{-p}^\dagger + \hat{a}_p^\dagger\hat{a}_p + \hat{a}_{-p}^\dagger\hat{a}_{-p}) \tag{11.102}$$

We can now write the Hamiltonian in the form

$$\hat{H} = \sum \left(\frac{N\alpha}{V} + \frac{p^2}{2m}\right)(\hat{a}_p^\dagger\hat{a}_p + \hat{a}_{-p}^\dagger\hat{a}_{-p}) + \frac{N\alpha}{V}\sum(\hat{a}_p\hat{a}_{-p} + \hat{a}_p^\dagger\hat{a}_{-p}^\dagger) + \frac{N^2\alpha}{2V} \tag{11.103}$$

The Hamiltonian is quadratic in the field operators and can be diagonalized by making a Bogoliubov transformation

$$\begin{aligned} \hat{a}_p &= u_p\hat{\alpha}_p + v_p\hat{\alpha}_{-p}^\dagger \\ \hat{a}_p^\dagger &= u_p\hat{\alpha}_p^\dagger + v_p\hat{\alpha}_{-p} \end{aligned} \tag{11.104}$$

The new operators $\hat{\alpha}_p^\dagger$ and $\hat{\alpha}_p$ satisfy the same commutation relations as $\hat{a}_p^\dagger$ and $\hat{a}_p$. This gives

$$u_p^2 - v_p^2 = 1 \tag{11.105}$$

Substitution of Eq. (11.104) in Eq. (11.103) gives

$$\begin{aligned} \hat{H} = &\sum_{p>0}\left\{\left(\frac{\hat{p}^2}{2m} + \frac{N\alpha}{V}\right)(u_p^2 + v_p^2) + \frac{2N\alpha}{V}u_pv_p\right\}(\hat{\alpha}_p^\dagger\alpha_p + \hat{\alpha}_{-p}^\dagger\hat{\alpha}_{-p}) \\ &+ \sum_{p>0}\left\{\left(\frac{\hat{p}^2}{2m} + \frac{N\alpha}{V}\right)2u_pv_p + \frac{N\alpha}{V}(u_p^2 + v_p^2)\right\}(\hat{\alpha}_p^\dagger\hat{\alpha}_{-p}^\dagger + \hat{\alpha}_p\hat{\alpha}_{-p}) \\ &+ \sum_{p>0}\left\{2\left(\frac{p^2}{2m} + \frac{N\alpha}{V}\right)v_p^2 + 2\frac{N\alpha}{V}u_pv_p\right\} + \frac{N^2\alpha}{2V} \end{aligned} \tag{11.106}$$

For the Hamiltonian to be diagonal, terms of the form $\hat{\alpha}_p^\dagger \hat{\alpha}_{-p}^\dagger$ should be absent. Hence we put

$$\left(\frac{\hat{p}^2}{2m} + \frac{N\alpha}{V}\right) 2u_p v_p + \frac{N\alpha}{V}(u_p^2 + v_p^2) = 0 \tag{11.107}$$

From Eqs. (11.105) and (11.107), the coefficients u_p and v_p can be obtained, and then inserting them in Eq. (11.106), we get

$$\hat{H} = E_0 + \sum_{p>0} \left[\left(\frac{\hat{p}^2}{2m} + \frac{N\alpha}{V}\right)^2 - \left(\frac{N\alpha}{V}\right)^2\right]^{1/2} (\hat{\alpha}_p^\dagger \hat{\alpha}_p + \hat{\alpha}_{-p}^\dagger \hat{\alpha}_{-p}) \tag{11.108}$$

where the zero point energy E_0 is equal to

$$E_0 = \frac{N^2\alpha}{2V} - \sum_{p>0} \left\{ \left(\frac{\vec{p}^2}{2m} + \frac{N\alpha}{V}\right) - \left[\left(\frac{\vec{p}^2}{2m} + \frac{N\alpha}{V}\right)^2 - \left(\frac{N\alpha}{V}\right)^2\right]^{1/2} \right\} \tag{11.109}$$

The weakly excited states of the system are superpositions of elementary excitations with energy spectrum

$$\epsilon_p = \left[\left(\frac{\vec{p}^2}{2m} + \frac{N\alpha}{V}\right)^2 - \left(\frac{N\alpha}{V}\right)^2\right]^{1/2} \tag{11.110}$$

For sufficiently small p

$$p^2 \ll 4m\frac{N\alpha}{V} \tag{11.111}$$

We have

$$\epsilon_p = p\sqrt{\frac{N\alpha}{mV}} \tag{11.112}$$

i.e., the elementary excitations are phonons with velocity equal to the velocity of sound and equal to $\sqrt{N\alpha/mV}$.

For the ground state energy E_0 as given by Eq. (11.109), the sum diverges. However, it was shown by Lee and Yang (*Phys. Rev.*, **105**, 1119, 1957) that finite results can be obtained by expressing the matrix element α in perturbation theory to second order. This gives

$$\alpha^{(2)} = \alpha - \frac{\alpha^2}{V}\sum \frac{m}{\vec{p}^2} \tag{11.113}$$

If we express α in terms of $\alpha^{(2)}$ and substitute in Eq. (11.109), we get

$$E_0 = \frac{N^2\alpha^{(2)}}{2V} - \sum \left\{ \left(\frac{\vec{p}^2}{2m} + \frac{N\alpha}{V}\right) - \left[\left(\frac{\vec{p}^2}{2m} + \frac{N\alpha}{V}\right)^2 - \left(\frac{N\alpha}{V}\right)^2\right]^{1/2} - \left(\frac{N\alpha}{V}\right)^2 \frac{m}{\vec{p}^2} \right\} \tag{11.114}$$

We can replace the sum in Eq. (11.114) by an integral and then performing the integration with α being replaced by scattering amplitude $a = m\alpha/\pi\hbar^2$, we come to the final result

$$E_0 = \frac{2\pi\hbar^2 N^2 a}{mV}\left(1 + \frac{128}{15\sqrt{\pi}}\sqrt{\frac{Na^3}{V}}\right) \tag{11.115}$$

The above represents the first-two terms in the expansion of E_0 in powers of $a(N/V)^{1/3}$ and is valid in the case of short-range forces, when the scattering amplitude is small in comparison with the interparticle distance.

The hard-sphere Bose gas can give only two phases, the gas phase and the condensed phase. Liquid He^4 has three phases, the gas, He I and He II. The hard-sphere gas has no attractive part and hence cannot form a bound state, which can be identified as a liquid. To obtain the three phases, an attractive interaction must be added to the hard-sphere interaction. Several forms of interaction potential between He atoms have been proposed. The most widely used is the Lannard–Jones potential.

In addition to the perturbation technique, variation technique has also been used for the evaluation of ground state energy of liquid helium. They mostly use the Bijl type of trial wave function

$$\psi(\vec{r}_1, \vec{r}_2, \ldots, \vec{r}_N) \simeq \prod_{1\le i<j\le N} w(|\vec{r}_i - \vec{r}_j|) \equiv \prod_{i<j} w(r_{ij}) \equiv \prod_{i<j} w_{ij} \tag{11.116}$$

The trial wave function w_{ij}, which depends explicitly on the interparticle separations, is defined to satisfy

$$\begin{aligned} w_{ij} &\to 0 \quad \text{if } r_{ij} \le a \\ w_{ij} &\to 1 \quad \text{if } r_{ij} \ge b \end{aligned} \tag{11.117}$$

and assume that the interaction energy can be expressed in the form of pair potentials

$$U(\vec{r}_1, \ldots, \vec{r}_N) = \sum_{1\le i\le j\le N} u(|\vec{r}_i - \vec{r}_j|) \equiv \sum_{i<j} u(r_{ij}) \equiv \sum_{i<j} u_{ij} \tag{11.118}$$

The use of Bijl type of wave functions has the advantage of the formal analogy of its expectation energy value with the configurational space integrals encountered in the classical equilibrium statistical mechanics. This analogy has been exploited to evaluate the expectation energy by techniques borrowed from the theory of classical fluids.

The techniques consist of the cluster expansion and the hyper-netted chain approximation and the use of Kirkwood superposition approximation relating the pair and triplet distribution functions. The quantum mechanical techniques of perturbation and variation have also been used. The expectation energy per particle for a Bose fluid of N identical particles (variation technique) is

$$\frac{E_0}{N} \le \frac{\rho}{2}\int d\vec{r}\left[-\frac{\hbar^2}{2m}\nabla^2 \ln w(r) + u(r)\right] g(r) \tag{11.119}$$

where $g(r)$ is the two-particle radial distribution function. Bounds on the distribution function have been developed and applied to obtain an upper bound on the ground state energy E_0. The radial distribution function has also been evaluated by the Monte Carlo method borrowed from the theory of classical fluids. The problem appears to have been very intelligently solved by Kalos, Levesque and Verlet (Phys. Rev. **A 9**, 2178, 1974) by using the 256 body systems with hard-sphere forces. They divided the interaction potential (LJ potential taken) into two parts (as done in classical liquids), one a repulsive part and the other part the remainder of the interaction. The repulsive part is replaced by suitable hard-spheres located at appropriate scattering length and the remainder attractive part is treated as perturbation. Their results agree with the experimental values.

EXERCISES

1. Show that the Bose–Einstein condensation takes place in liquid helium and estimate its value.
2. What are the salient features of the two-fluid theory of liquid He^4? How does it explain the phenomenon of second sound?
3. Discuss Landau theory of liquid He^4 and show how it improves the agreement of the velocity of the second sound with the experimental observation?
4. Discuss Feynman theory of liquid He^4. How does it improve upon Landau theory?

12 Non-Equilibrium States

BOLTZMANN TRANSPORT EQUATION

Till now we have dealt with systems in equilibrium (macroscopic state characterized by a few parameters). If a system is observed for a sufficiently long interval of time, it remains in an equilibrium state for a greater part of it. This is so because a system in non-equilibrium state has a general tendency to approach the state of equilibrium. The state of equilibrium is brought about by interactions between the individual particles, which can be understood only on microscopic scale. For a system in non-equilibrium state, we therefore need a distribution function (probability density in phase space) which is time-dependent, i.e. a function of position, momentum (or velocity) as well as time and also the time development of the distribution function. Boltzmann transport equation is the first step in this direction.

Let us consider a dilute gas, so that only two-particle interactions are important. The number of j particles in phase volume $d\vec{r}\, d\vec{v}_j$ about the point $\vec{r}$, $\vec{v}_j$ is given by $f_j(\vec{r}, \vec{v}_j, t)\, d\vec{r}\, d\vec{v}_j$, where $f_j(\vec{r}, \vec{v}_j, t)$ is the distribution function.

f_j varies with time in two independent ways:

(i) f_j may vary due to the motion of particles from one region to another as well as due to acceleration of particles caused by external forces $\vec{X}_j$ during the motion. This change is called *drift variation* and the rate of change due to drift is given by

$$\left(\frac{df_j}{dt}\right)_d = \left[\frac{\partial f_j}{\partial t} + \vec{v}_j \cdot \nabla_r f_j + \frac{\vec{X}_j}{m} \cdot \nabla_{v_j} f_j\right] d\vec{r}\, d\vec{v}_j\, dt \tag{12.1}$$

(ii) f_j may vary because of discontinuous changes accompanying collisions.

The number of j particles lost from the range $(\vec{r}, \vec{r}+d\vec{r})$, $(\vec{v}_j, \vec{v}_j + d\vec{v}_j)$ due to collisions with i-particles in the time interval t and $t + dt$ is $A_{ji}^{-}\, d\vec{r}\, d\vec{v}_j\, dt$, where A_{ji}^{-} is probability per unit time per unit volume of losing the range.

Similarly, the number of j-particles that joins the group in the range because of collisions with i-particles in the time interval t and $t + dt$ is $A_{ji}^{+}\, d\vec{r}\, d\vec{v}_j\, dt$, where A_{ji}^{+} is the probability per unit time per unit volume of joining the group.

Then, the rate of change due to collisions is

$$\left(\frac{df_j}{dt}\right)_c = \sum_i (A_{ji}^{+} - A_{ji}^{-})\, d\vec{r}\, d\vec{v}_j\, dt \tag{12.2}$$

The condition for equilibrium gives

$$\frac{\partial f_j}{\partial t} + \vec{v}_j \cdot \nabla_r f_j + \frac{\vec{X}_j}{m} \cdot \nabla_{v_j} f_j = \sum_i (A_{ji}^{+} - A_{ji}^{-}) \tag{12.3}$$

We now determine the explicit form of the collision terms. The probability that a particle of j-type located at $\vec{r}$ with velocity $\vec{v}_j$ will collide with an i-particle in the time interval dt with impact parameter in the range b and $b + db$ is calculated in the following way.

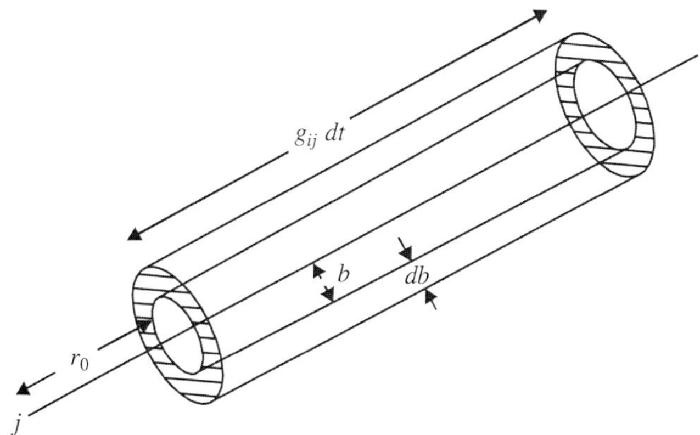

Figure 12.1 Schematic diagram for collision of molecules.

Let the j-particle be considered fixed, then the i-particle approaches it with a relative velocity $(\vec{v}_i - \vec{v}_j) = \vec{g}_{ij}$. Any i-particle in a cylindrical shell of length $g_{ij}\, dt$ beyond the range r_0 of the interparticle potential will collide with the fixed j-particle in the time interval dt. The number of particles in the cylindrical shell in the velocity range $\vec{v}_i$ and $\vec{v}_i + d\vec{v}_i$ is

$$f_i(\vec{r}, \vec{v}_i, t)\, 2\pi b\, db\, g_{ij}\, d\vec{v}_i\, dt$$

Hence the number of collisions that will occur with the fixed j-particles is

$$2\pi\, dt \int f_i(\vec{r}, \vec{v}_i, t) b\, db\, g_{ij}\, d\vec{v}_i$$

The number of j-particles located in phase space volume element $d\vec{r}\, d\vec{v}_j$ is

$$f_j(\vec{r}, \vec{v}_j, t)\, d\vec{r}\, d\vec{v}_j$$

This gives

$$A_{ji}^{-}\, d\vec{r}\, d\vec{v}_j\, dt = 2\pi\, d\vec{r}\, d\vec{v}_j\, dt \iint f_j(r, v_j, t)\, f_i(\vec{r}, \vec{v}_i, t)\, b\, db\, g_{ij}\, d\vec{v}_i \tag{12.4}$$

which simplifies to

$$A_{ji}^{-} = 2\pi \iint f_j f_i\, b\, db\, g_{ij}\, d\vec{v}_i \tag{12.5}$$

We can calculate A_{ji}^{+} in a similar way. We are now interested in inverse collisions of the type $\{\vec{v}_j', v_i'\} \rightarrow \{\vec{v}_j, \vec{v}_i\}$. This will give

$$A_{ji}^{+}\, d\vec{r}\, d\vec{v}_j\, dt = 2\pi\, d\vec{r}\, d\vec{v}_j'\, dt \iint f_i(\vec{r}, \vec{v}_i', t)\, f_j(\vec{r}, \vec{v}_j', t)\, b'\, db'\, g_{ij}'\, d\vec{v}_i' \tag{12.6}$$

The primes indicate the quantities before collision, which go to b, $\vec{v}_i$ and $\vec{v}_j$.

From Liouville theorem, we have

$$d\vec{r}\, d\vec{v}_i\, b\, db\, d\vec{v}_j\, g_{ij}\, dt = d\vec{r}\, d\vec{v}_i'\, b'\, db'\, g_{ij}'\, d\vec{v}_j'\, dt \tag{12.7}$$

Substitution from Eq. (12.7) into Eq. (12.6) gives

$$\begin{aligned} A_{ji}^{+} &= 2\pi \iint f_i(\vec{r}, v_i', t)\, f_j(\vec{r}, v_j', t)\, b\, db\, g_{ij}\, d\vec{v}_i \\ &= 2\pi \iint f_i' f_j'\, b\, db\, g_{ij}\, d\vec{v}_i \end{aligned} \tag{12.8}$$

The primes on f's indicate that their velocity arguments are primed.

Substitution from Eqs. (12.5) and (12.8) into Eq. (12.3) gives

$$\frac{\partial f_j}{\partial t} + \vec{v}_j \cdot \nabla_r f_j + \frac{\vec{X}_j}{m} \cdot \nabla_v f_j = 2\pi \iint [f_i' f_j' - f_i f_j]\, b\, db\, g_{ij}\, d\vec{v}_i \tag{12.9}$$

This is Boltzmann transport equation; the functions f_i' and f_j' depend upon post collision velocities $\vec{v}_i'$ and $\vec{v}_j'$, which are related to precollision velocities $\vec{v}_i$, $\vec{v}_j$ according to the equations of the collision problem. The Boltzmann equation is an integro-differential equation for f giving its time development.

The above calculation of $(df/dt)_c$ is based on the assumption of molecular chaos (the velocity and the position of particles are uncorrelated), which makes it possible to express it in terms of f itself. Otherwise, $(df/dt)_c$ will involve two-particle correlation function which is independent of f, and in place of Eq. (12.9), we will have an equation relating f to two-particle correlation function (see Bogoliubov, N.N., in Boer, J. de and Uhlenbeck, E.G., *Studies in Statistical Mechanics*, North-Holland Publishing Co., 1962, Vol. I).

BOLTZMANN *H*-THEOREM

Consider the function

$$H(t) = \iint f(\vec{r}, \vec{v}, t) \ln f(r, \vec{v}, t) \, d\vec{r} \, d\vec{v} \tag{12.10}$$

known as Boltzmann *H*-function. Differentiating with respect to time, we get

$$\frac{dH}{dt} = \iint \frac{\partial f}{\partial t} \ln f \, d\vec{r} \, d\vec{v} + \iint \frac{\partial f}{\partial t} \, d\vec{r} \, d\vec{v} \tag{12.11}$$

The second term vanishes because of normalization condition:

$$\iint \frac{\partial f}{\partial t} \, d\vec{r} \, d\vec{v} = \frac{d}{dt} \iint f \, d\vec{r} \, d\vec{v} = \frac{dN}{dt} = 0$$

(The number of particles in the system is conserved.)

This gives

$$\frac{dH}{dt} = \iint \frac{\partial f}{\partial t} \ln f \, d\vec{r} \, d\vec{v} \tag{12.12}$$

Multiplying Boltzmann equation (12.9) by $\ln f$ and integrating over $d\vec{r} \, d\vec{v}$, we get

$$\iint \frac{\partial f}{\partial t} \ln f \, d\vec{r} \, d\vec{v} = -\iint (\ln f) \, \vec{v} \cdot \nabla_r f \, d\vec{r} \, d\vec{v} - \iint (\ln f) \frac{\vec{X}}{m} \cdot \nabla_v f \, d\vec{r} \, d\vec{v}$$

$$+ 2\pi \iint\iint (\ln f)(f'f_1' - ff_1) \, b \, db \, g \, d\vec{v}_1 \, d\vec{r} \, d\vec{v} \tag{12.13}$$

(We are using no subscript for j and 1 for i.)

The first-two integrals on right-hand side vanish since f vanishes at the walls of container and as $\vec{v} \to \pm\infty$. This gives

$$\frac{dH}{dt} = 2\pi \int \cdots \int \ln f (f'f_1' - ff_1) \, b \, db \, g \, d\vec{v}_1 \, d\vec{r} \, d\vec{v} \tag{12.14}$$

Interchange of $\vec{v}_1$ and $\vec{v}$ gives an alternative form

$$\frac{dH}{dt} = 2\pi \int \cdots \int \ln f_1 (f'f_1' - ff_1) \, b \, db \, g \, d\vec{v}_1 \, d\vec{r} \, d\vec{v} \tag{12.15}$$

Hence by adding the two and dividing by half, we can write

$$\frac{dH}{dt} = 2\pi \int \cdots \int \frac{1}{2} \ln ff_1 \left(f'f_1' - ff_1\right) b \, db \, g \, d\vec{v}_1 \, d\vec{r} \, d\vec{v} \tag{12.16}$$

For every collision, there corresponds an inverse collision. Hence $\{\vec{v}, \vec{v}_1\} \to \{\vec{v}', \vec{v}_1'\}$, i.e. interchanging the primed and unprimed terms and using Eq. (12.16), we have

$$\frac{dH}{dt} = 2\pi \int \cdots \int \frac{1}{2} \ln f'f_1' (ff_1 - f'f_1') \, b \, db \, g \, d\vec{v}_1 \, d\vec{r} \, d\vec{v} \tag{12.17}$$

Adding Eqs. (12.16) and (12.17) and dividing by two, we get

$$\frac{dH}{dt} = 2\pi \int \cdots \int \frac{1}{4} \ln \frac{ff_1}{f'f_1'} (f'f_1' - ff_1)\, b\, db\, g\, d\vec{v}_1\, d\vec{r}\, d\vec{v} \tag{12.18}$$

The integrand on the right-hand side is always negative; for $f'f_1' > ff_1$, $\ln (ff_1/f'f_1')$ is negative and for $f'f_1' < ff_1$, $(f'f_1' - ff_1)$ is negative. Hence we get the result

$$\frac{dH}{dt} = \leq 0 \tag{12.19}$$

The equality sign holds only if $ff_1 = f'f_1'$. This means that the distribution function is independent of time and corresponds to equilibrium or steady-state solution with $ff_1 = f'f_1' \cdot (dH/dt) < 0$ corresponds to non-stationary solution in which the distribution function is an explicit function of time. Farther the system is from equilibrium, more likely $H(t)$ is to decrease, Ehrenfest and Smoluchowski later have shown that on the average, the H-function decreases with time to the equilibrium value and remains near it once it gets there. We shall show that the steady-state solution corresponds to Maxwellian distribution of velocities.

In steady state

$$ff_1 = f'f_1'$$

or

$$\ln f + \ln f_1 = \ln f' + \ln f_1' \tag{12.20}$$

i.e. $\ln f$ is a summational invariant. The only summational invariants of a spherically symmetric two particle collisions are the mass, the momentum and the kinetic energy. Hence $\ln f$ must be a linear combination of these quantities, i.e.

$$\ln f = \alpha m + \vec{\beta} \cdot (m\vec{v}) - \gamma \frac{mv^2}{2} \tag{12.21}$$

where α, β and γ are constants.

This can be written in the form

$$\ln f = \ln C - A(\vec{v} - \vec{v}_0)^2 \tag{12.22}$$

where A and C are constants and $\vec{v}_0 = \langle \vec{v} \rangle$

This gives

$$f = Ce^{-Av^2} \tag{12.23}$$

Since $\vec{v}_0 = \langle \vec{v} \rangle$ can be put equal to zero. Normalization gives

$$\int_0^\infty f\, d\vec{v} = \frac{N}{V} = \rho$$

whence

$$\rho = C \int_0^\infty e^{-Av^2}\, d\vec{v} = C \left(\frac{\pi}{A} \right)^{3/2} \tag{12.24}$$

in which we put $v^2 = v_x^2 + v_y^2 + v_z^2$ and $d\vec{v} = dv_x\, dv_y\, dv_z$.

This gives

$$C = \rho \left(\frac{A}{\pi} \right)^{3/2} \tag{12.25}$$

For average energy

$$\begin{aligned} \langle \in \rangle &= \frac{C}{\rho} \int_0^\infty \frac{1}{2} m v^2 e^{-Av^2} d\vec{v} \\ &= \frac{Cm}{2\rho} 4\pi \int_0^\infty v^4 e^{-Av^2} dv \\ &= \frac{3}{4} \frac{m}{A} \end{aligned} \tag{12.26}$$

If we identify $\langle \in \rangle$ by 3/2 kT, we get

$$A = \frac{m}{2kT} \tag{12.27}$$

Substitution from Eqs. (12.25) and (12.27) into Eq. (12.23) gives

$$f = \rho \left(\frac{m}{2\pi kT} \right)^{3/2} \exp\left(-\frac{mv^2}{2kT} \right) \equiv \rho \left(\frac{m}{2\pi kT} \right)^{3/2} \exp\left(-\frac{m(\vec{v} - \vec{v}_0)^2}{2kT} \right) \tag{12.28}$$

which is the classical expression for Maxwellian distribution of velocities.

To relate H-function and entropy, we recall the thermodynamic inequality

$$\frac{dS}{dt} \geq 0 \tag{12.29}$$

The entropy of classical system of N-identical particles in thermodnamic equilibrium is

$$S = -k \int f \ln f \, dp \, dq \tag{12.30}$$

where f is the probability density for a canonical distribution. Since the particles of an ideal gas are statistically independent, we have

$$f(p,q) = \prod_{i=1}^{N} f_i(p_i, q_i) \tag{12.31}$$

where $f_i(p_i, q_i)$ is the single particle probability density, given by

$$f_i(p_i, q_i) = \int f(p, q) \, dp_1 \cdots dp_{i-1} \, dp_{i+1} \cdots dp_N \, dq_1 \cdots dq_{i-1} dq_{i+1} \cdots dq_N \tag{12.32}$$

This gives

$$S = -k \sum_{j=1}^{N} \int f_j \ln f_j \, dp_j \, dq_j = -Nk \int f_j \ln f_j \, dp_j \, dq_j$$

Since the number of particles in volume $d\vec{r}\,d\vec{v}_j$ at $\vec{r}$ and $\vec{v}_j$ is given by $f(\vec{r},\vec{v}_j,t)\,d\vec{r}\,d\vec{v}_j$ and is also equal in terms of single particle probability density to $Nf_j\,d\vec{r}\,d\vec{p}_j = Nm^3 f_j\,d\vec{r}\,d\vec{v}_j$, we have

$$S = -k \iint f(\vec{r},\vec{v},t) \ln f(\vec{r},\vec{v},t)\, d\vec{r}\, d\vec{v} + Nk \ln(Nm^3) \tag{12.33}$$

This gives

$$S - S_0 = -kH \tag{12.34}$$

where

$$S_0 = Nk \ln (Nm^3) \tag{12.35}$$

This means that but for an additive constant, entropy is proportional to minus the Boltzmann H-function.

THEORY OF BROWNIAN MOTION

Robert Brown, an English botanist, observed in 1827, through a microscope, grains of pollen suspended in water. He found that these grains were taking part in a much-agitated dance, which continued without any change in vigour. This movement is referred to as *Brownian motion*. The following facts were observed about it.

(i) The suspended particles undergo random motion.

(ii) The motion is continuous and eternal. Brownian motion of liquids enclosed in cavities of some varieties of quartz have been observed, which had been there for thousands of years.

(iii) The motion is slower in more viscous liquids.

(iv) The motion is quicker if the particle size is smaller.

Different explanations for it were proposed from time to time, only to be discarded after careful experiments. Einstein, in 1905, suggested that fluctuations in collisions between the suspended pollen particles and the molecules of the surrounding fluid were responsible for it. Mathematical details were worked out by Einstein and Smoluchowski and were experimentally verified by Perrin in 1907.

The problem can be easily treated by a method due to Langevin (1908), which arrives at the same result in a simpler way. The Langevin method is also applicable to other random processes.

The equation of motion of a particle of mass m with a force $F(t)$ acting upon it is

$$m\frac{d^2x}{dt^2} = F(t)$$

We assume that the force acting on a particle can be split into two parts. One is due to viscous drag exerted by the liquid on the particle and is proportional to the velocity. The other part, a fluctuating force $\beta(t)$ on the particle, is exerted by the continuous impact of the molecules of the liquid. This force varies extremely rapidly in comparison with the changes in velocity of the particle. Hence it can be assumed that no correlation between $\beta(t + \Delta t)$ and $\beta(t)$ exists. The Langevin equation is the simple equation of motion under the above conditions

$$m\frac{d^2x}{dt^2} + \delta\frac{dx}{dt} - \beta(t) = 0 \tag{12.36}$$

where δ is the damping coefficient due to viscosity and $\beta(t)$ is the fluctuating force.

Multiplying througout by x and recalling the identity

$$x\,\frac{d^2x}{dt^2} = \frac{1}{2}\,\frac{d^2}{dt^2}(x^2) - \left(\frac{dx}{dt}\right)^2 \tag{12.37}$$

we get

$$\frac{1}{2}m\,\frac{d^2}{dt^2}(x^2) - m\left(\frac{dx}{dt}\right)^2 + \frac{1}{2}\delta\,\frac{d}{dt}(x^2) - x\beta(t) = 0 \tag{12.38}$$

Taking the time average of Eq. (12.38), and also noting that $\langle x\beta(t)\rangle = 0$ since $\beta(t)$ is randomly fluctuating and will have on the average positive and negative values, and using the relation from equipartition theorem

$$\frac{1}{2}m\left\langle\left(\frac{dx}{dt}\right)^2\right\rangle = \frac{1}{2}\,kT$$

we get

$$\frac{1}{2}m\,\frac{d\alpha}{dt} - kT + \frac{1}{2}\,\delta\alpha = 0 \tag{12.39}$$

where $\alpha = d/dt\,\langle x^2\rangle$.

Thus Eq. (12.39) simplifies to

$$\frac{m}{\delta}\frac{d\alpha}{dt} + \alpha = \frac{2kT}{\delta}$$

On integration (using an integrating factor $e^{(\delta/m)t}$), we get

$$\alpha = \frac{d}{dt}\,\langle x^2\rangle = \frac{2kT}{\delta} + C\,e^{-(\delta/m)t} \tag{12.40}$$

where C is a constant of integration.

The exponential gives the transient state. From Stokes law $\delta = 6\pi a\eta$, a is the radius of the suspended particle and η is the coefficient of viscosity. For a particle size of the order of 10^{-4} cm, we see that the transient term becomes zero for any appreciable value of time. Dropping out this term, we can write

$$\alpha = \frac{d}{dt}\langle x^2\rangle = \frac{2kT}{\delta} \tag{12.41}$$

Integrating again, we have

$$\langle x^2 \rangle = \frac{2kT}{\delta} t = \frac{kT}{3\pi a\eta} t \tag{12.42}$$

which is the equation originally derived by Einstein.

If we use the full solution of Eq. (12.40) with the initial conditions at $t = 0$, $x = 0$, we get

$$\langle x^2 \rangle = \frac{2kT}{\delta}\left[t - \frac{m}{\delta}\left(1 - e^{-(\delta/m)t}\right)\right] \tag{12.43}$$

From this we see that for $t \ll m/\delta$, $\langle x^2 \rangle = (kT/m)t^2$, which corresponds to the particle velocity $(kT/m)^{1/2}$ that is same as the thermal velocity. For $t \gg m/\delta$, we have $\langle x^2 \rangle \simeq (2kT/\delta)t$, the Brownian movement sets in.

RANDOM WALK IN ONE DIMENSION

Let us consider the problem of a particle making n jumps (each of length l) per unit time, with equal probability along the positive or negative direction of x. We want to find the probability that the particle will travel a distance x in time interval t. The original problem of a drunken sailor taking the random walk (moving forward or backward in steps), reaching his home in a given time, was solved by Markov. For this reason, it is also called *Markov process.*

Let the particle make N jumps (steps) in a time t such that the number of jumps per second $n = N/t$. The probability $P(m)$ of a distance traversed $x = ml$ is equal to the probability that it takes $(N + m)/2$ jumps of positive type and $(N - m)/2$ jumps of the negative type.

The probability of the jump being in the either direction is 1/2. The probability that N jumps have a specified sequence is $(1/2)^N$. Hence

$$P(m) = C\left(\frac{1}{2}\right)^N \times \text{Number of distinct sequences to take } m \text{ excess positive jumps} \tag{12.44}$$

where C is a normalizing constant.

The N jumps can be permuted in $N!$ ways amongst themselves. Of them, the $(N + m)/2$ positive jumps can be permuted in $[(N + m)/2]!$ ways amongst themselves and the $(N - m)/2$ negative jumps can be permuted in $[(N - m)/2]!$ ways amongst themselves. This gives

$$P(m) = C\left(\frac{1}{2}\right)^N \frac{N!}{[(N+m)/2]!\,[(N-m)/2]!} \tag{12.45}$$

Since N is large and $m << N$, we use the Stirling approximation in its more exact form

$$\ln n! \simeq \frac{1}{2}\ln 2\pi + \left(n + \frac{1}{2}\right)\ln n - n \tag{12.46}$$

This gives

$$\ln P(m) = \ln C - N\ln 2 - \frac{1}{2}\ln 2\pi + \left(N+\frac{1}{2}\right)\ln N$$

$$-\frac{1}{2}(N+m+1)\ln\left[\frac{1}{2}(N+m)\right] - \frac{1}{2}(N-m+1)\ln\left[\frac{1}{2}(N-m)\right]$$

$$= \ln C - N\ln 2 - \frac{1}{2}\ln 2\pi + \left(N+\frac{1}{2}\right)\ln N$$

$$-\frac{1}{2}(N+m+1)\ln\left[\frac{N}{2}\left(1+\frac{m}{N}\right)\right] - \frac{1}{2}(N-m+1)\ln\left[\frac{N}{2}\left(1-\frac{m}{N}\right)\right]$$

For $m \ll N$, we use the approximation

$$\ln\left(1 \pm \frac{m}{N}\right) \simeq \pm\frac{m}{N} - \frac{m^2}{2N^2}$$

to get

$$\ln P(m) = \ln C + \ln 2 - \frac{1}{2}\ln 2\pi - \frac{1}{2}\ln N - \frac{m^2}{2N}$$

whence

$$P(m) = C\sqrt{\frac{2}{\pi N}}\, e^{-m^2/2N} \tag{12.47}$$

From normalization condition

$$\int P(m)\, dm = 1$$

we get

$$C = \frac{1}{2}$$

Substitution gives

$$P(m)\, dm = \frac{1}{\sqrt{2\pi N}}\, e^{-m^2/2N}\, dm \tag{12.48}$$

Using $x = ml$ and $t = N/n$, we get the probability of the particle reaching a distance x in time t as

$$P(x)\, dx = \left(\frac{1}{2\pi l^2 nt}\right)^{1/2} e^{-x^2/2l^2nt} \tag{12.49}$$

It is seen that the probability has a Gaussian distribution and the width of the peak spreads as a square root of time.

The mean value $\langle x\rangle = 0$ is easily seen. The mean square displacement

$$\langle x^2\rangle = \int_{-\infty}^{\infty} x^2 P(x)\, dx = \frac{1}{(2\pi l^2 nt)^{1/2}}\int x^2\, e^{-x^2/2l^2nt}\, dx$$

$$= l^2 nt \tag{12.50}$$

DIFFUSION

The random walk process is used as a model for the process of diffusion. The particle current density $\vec{J}(\vec{r},t)$ is related to the probable particle concentration by Fick law as

$$\vec{J}(\vec{r},t) = -D \,\text{grad}\, \rho(\vec{r},t) \tag{12.51}$$

where D is the coefficient of diffusion. Combining this with the equation of continuity for the system

$$\frac{\partial \rho(\vec{r},t)}{\partial t} + \text{div}\, \vec{J}(\vec{r},t) = 0 \tag{12.52}$$

we obtain the diffusion equation

$$\frac{\partial \rho(\vec{r},t)}{\partial t} = D\nabla^2 \rho(\vec{r},t) \tag{12.53}$$

We take the one-dimensional case for simplicity

$$\frac{\partial \rho(x,t)}{\partial t} = D\frac{\partial^2 \rho(x,t)}{\partial x^2} \tag{12.54}$$

The appropriate solution of the equation is

$$\rho(x,t) = \frac{N}{\sqrt{4\pi Dt}} \exp\left(-\frac{x^2}{4Dt}\right) \tag{12.55}$$

The mean value of the spread is

$$\langle x^2 \rangle = \frac{1}{N}\frac{N}{\sqrt{4\pi Dt}} \int_{-\infty}^{\infty} x^2 \exp\left(-\frac{x^2}{4Dt}\right) dx = 2Dt \tag{12.56}$$

We can tackle the problem by random walk method. Let the number of particles be N_0 at $x = 0$, $t = 0$, making random jumps of step size l, and let n be the number of jumps made per unit time. Then the probability of the jumps being in the x-direction (positive or negative) is $n/3$. The probable particle concentration at the position x and time t is

$$\begin{aligned} \rho(x,\ t) &= N_0 P(x)\ dx \\ &= N_0 \left(\frac{3}{2\pi l^2 nt}\right) e^{-3x^2/2l^2nt} \end{aligned} \tag{12.57}$$

and the mean value of the spread [replace n by $n/3$ in Eq. (12.50)] is

$$\langle x^2 \rangle = \frac{1}{3} l^2 nt \tag{12.58}$$

Comparing Eqs. (12.56) and (12.58), we get

$$D = \frac{1}{6} l^2 n \tag{12.59}$$

Comparing Eq. (12.56) with Eq. (12.42), we get

$$D = \frac{kT}{\delta} = kT\mu \tag{12.60}$$

The reciprocal of δ is the drift velocity per unit applied force and is called the mobility of the particle, denoted by μ.

The relation connecting the coefficient of diffusion with the mobility of the particles is due to Einstein.

TIME-CORRELATION FUNCTION

Till now we have been dealing with the estimation of the mean square values of fluctuations. But the time dependence of the fluctuations is also of considerable interest. Green and Kubo have shown that the phenomenological coefficients describing many transport processes and time-dependent phenomena in general could be written as integrals over a certain type of function, called a *time-correlation function*. The advantage of time-correlation formalism is that the expressions derived for the transport coefficients are general in the sense that they are not limited by the choice of any model or any density region. Further, they lead to exact equations whose solutions are very difficult but enable us to make approximations. The very nature of time-correlation function leads nicely to this procedure. A time-dependent perturbation produces a time-dependent response which can be Fourier analyzed to give the spectral density of the fluctuating quantity. Further, the time-dependent response is linearly related to the time-dependent perturbation, leading to the linear response theory.

The time-correlation function of any two dynamical variables $A(t)$ and $B(t)$ is defined as the ensemble average

$$C_{AB}(\tau) = \langle\langle A(t+\tau)\, B(t)\rangle\rangle = \left\langle \underset{T\to\infty}{\mathrm{Lt}} \frac{1}{T}\int_0^\infty A(t+\tau)\, B(t)\, dt \right\rangle \tag{12.61}$$

The first average stands for time average and the second stands for ensemble average.

If $A = B$, then

$$C_A(\tau) = \langle\langle A(t+\tau)\, A(t)\rangle\rangle = \left\langle \underset{T\to\infty}{\mathrm{Lt}} \frac{1}{T}\int_0^\infty A(t+\tau)\, A(t)\, dt \right\rangle \tag{12.62}$$

$C_A(\tau)$ is called time *auto-correlation function*.

For small τ, it is expected that $A(t + \tau)$ be nearly equal to $A(t)$ and we can write

$$C_A(0) = \langle\langle A^2\rangle\rangle \tag{12.63}$$

As $\tau \to \infty$, $C_A(\tau)$ must tend to zero because fluctuations are random and there will be no correlation between $A(t + \tau)$ and $A(t)$. For a rapidly fluctuating function, $C_A(\tau)$ will approach zero for relatively smaller values of τ.

FOURIER ANALYSIS OF FLUCTUATION: WIENER–KHINCHIN THEOREM

The time scales of fluctuations can be analyzed by Fourier decomposition. Let $A(t)$ be a randomly fluctuating quantity with time. If we observe for a long interval of time T such that $A(0) = A(T) = A$, then the fluctuations can be expressed as a Fourier series

$$A(t) = \sum_{r=0}^{\infty} \left[a_r \cos\frac{2\pi rt}{T} + b_r \sin\frac{2\pi rt}{T} \right] \tag{12.64}$$

where

$$a_r = \frac{2}{T}\int_0^T A(t)\cos\frac{2\pi rt}{T}\,dt$$

$$b_r = \frac{2}{T}\int_0^T A(t)\sin\frac{2\pi rt}{T}\,dt \tag{12.65}$$

The time average is

$$\begin{aligned}\langle A(t+\tau)\,A(t)\rangle = \underset{T\to\infty}{\mathrm{Lt}}\,\frac{1}{T}\Bigg[\int_0^T \left(\sum_r a_r\cos\frac{2\pi r(t+\tau)}{T} + \sum_r b_r\sin\frac{2\pi r(t+\tau)}{T}\right) \\ \left(\sum_r a_r\cos\frac{2\pi rt}{T} + \sum_r b_r\sin\frac{2\pi rt}{T}\right)dt\Bigg]\end{aligned} \tag{12.66}$$

Using

$$\cos\frac{2\pi r(t+\tau)}{T} = \cos\frac{2\pi rt}{T}\cos\frac{2\pi r\tau}{T} - \sin\frac{2\pi rt}{T}\sin\frac{2\pi r\tau}{T}$$

$$\sin\frac{2\pi r(t+\tau)}{T} = \sin\frac{2\pi rt}{T}\cos\frac{2\pi r\tau}{T} + \cos\frac{2\pi rt}{T}\sin\frac{2\pi r\tau}{T}$$

$$\int_0^T \cos^2\frac{2\pi rt}{T}\,dt = \frac{T}{2} = \int_0^T \sin^2\frac{2\pi rt}{T}\,dt$$

$$\int_0^T \cos\frac{2\pi rt}{T}\sin\frac{2\pi rt}{T}\,dt = 0$$

we get

$$C(t) = \frac{1}{2}\sum_{r=0}^{\infty}[\langle a_r^2\rangle + \langle b_r^2\rangle]\cos\frac{2\pi r\tau}{T} \tag{12.67}$$

Each value of r corresponds to a frequency $\nu_r = r/T$ and the amount of power contained in this frequency is $1/2\,[\langle a_r^2\rangle + \langle b_r^2\rangle]$ and is called the spectral density of the randomly fluctuating quantity, i.e.

$$G(\nu) \leftrightarrow \frac{1}{2}[\langle a_r^2 \rangle + \langle b_r^2 \rangle] \tag{12.68}$$

where $G(\nu)$ is the spectral density.

If we go from discrete to continuous distribution of frequencies, Eq. (12.67) with the aid of Eq. (12.68) is written as

$$C(\tau) = \int_0^{\infty} G(\nu) \cos 2\pi\nu\tau \, d\nu \tag{12.69}$$

This gives the relation between the time-correlation function and the power spectrum of the fluctuations, which is a Fourier cosine transform. The inverse relation is

$$G(\nu) = 4 \int_0^{\infty} C(\tau) \cos 2\pi\nu\tau \, d\tau \tag{12.70}$$

The pair of relations given by Eqs. (12.69) and (12.70) is called *Wiener–Khinchin theorem* in the study of fluctuations.

The pair of relations given by Eqs. (12.69) and (12.70) can be written in complex form as

$$C(\tau) = \int_{-\infty}^{\infty} G(\omega) e^{i\omega\tau} \, d\omega \tag{12.71}$$

$$G(\omega) = \frac{1}{2\pi} \int_{-\infty}^{\infty} C(\tau) \, e^{-i\omega\tau} \, d\tau \tag{12.71a}$$

The Wiener–Khinchin theorem allows us to guess how rapidly random variations will relax to equilibrium. If we allow the fluctuations to decay exponentially with time with a relaxation time τ_c, we have

$$C(\tau) = C(0) \, e^{-\tau/\tau_c} \tag{12.72}$$

from which we get

$$\begin{aligned} G(\nu) &= 4 \int_0^{\infty} C(0) \, e^{-\tau/\tau_c} \cos 2\pi\nu\tau \, d\tau \\ &= \frac{4\tau_c C(0)}{1 + 4\pi^2 \nu^2 \tau_c^2}, \quad G(0) = 4\tau_c C(0) \end{aligned} \tag{12.73}$$

where we use

$$\int_0^{\infty} e^{-at} \cos mt \, dt = \frac{a}{a^2 + m^2}$$

A plot of $G(\nu)/G(0)$ against $\nu/2\pi\tau_c$ is shown in Fig. 12.2. We find that $G(\nu)$ is almost constant for $\nu/2\pi\tau_c \ll 1$ and rapidly goes to zero for frequencies higher than $1/2\pi\tau_c$.

The Wiener–Khinchin theorem is useful in analyzing the spectral distribution of energy of light scattered by fluctuations from a medium. It is also very useful in the study of fluctuations in electrical and electronic systems (time-dependent noise problem).

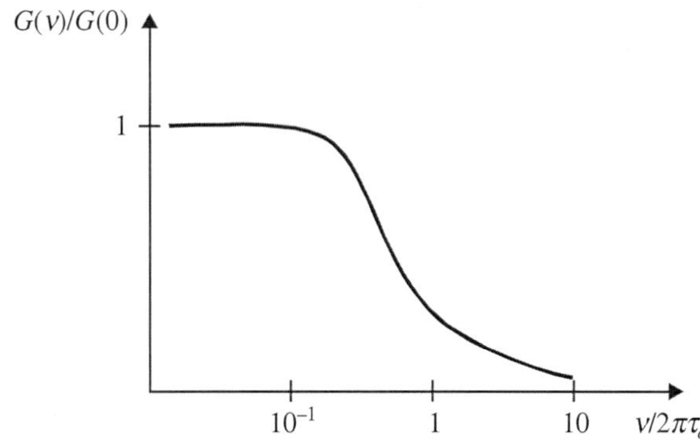

Figure 12.2 Power spectrum of fluctuations for exponentially decaying correlations.

FOKKER–PLANCK EQUATION

We examine the time pertaining behaviour of a system subjected to fluctuating forces and obtain the Fokker–Planck equation. It is an equation of motion for the distribution f. Let $f(v, t)$ be the probability that a particle has velocity v at time t. For simplicity, we consider the one-dimensional case only. We further assume that $\phi(v, \xi)\, d\xi$ is the probability of finding the particle velocity between $v + \xi$ and $v + \xi + d\xi$ after a time interval τ, if its initial velocity was v. Then,

$$f(v, t+\tau) = \int_{-\infty}^{\infty} f(v-\xi, t)\, \phi(v-\xi, \xi)\, d\xi \tag{12.74}$$

For small time τ, velocities change only by small amount, i.e. $\xi \ll v$. Hence we make a Taylor expansion of $f(v - \xi, t)$ and $\phi(v - \xi, \xi)$ and retain terms up to ξ^2 only

$$f(v-\xi, t) = f(v, t) - \xi \frac{\partial f(v,t)}{\partial v} + \frac{\xi^2}{2}\frac{\partial^2 f(v,t)}{\partial v^2} \tag{12.75}$$

$$\phi(v-\xi, \xi) = \phi(v, \xi) - \xi \frac{\partial \phi(v,\xi)}{\partial v} + \frac{\xi^2}{2}\frac{\partial^2 \phi(v,\xi)}{\partial v^2} \tag{12.76}$$

and a similar expansion of $f(v, t + \tau)$ retaining only terms up to τ only

$$f(v, t+\tau) = f(v,t) + \tau \frac{\partial f(v,t)}{\partial t} \tag{12.77}$$

Substituting from Eqs. (12.75), (12.76) and (12.77) into Eq. (12.74) and retaining terms only up to ξ^2, we get

$$f(v,t) + \tau\frac{\partial f(v,t)}{\partial t} = \int \left\{ f(v,t)\, \phi(v,\xi)\, d\xi - \xi \left[f(v,t)\frac{\partial \phi(v,\xi)}{\partial v} + \frac{\partial f(v,t)}{\partial v}\phi(v,\xi) \right] \right.$$

$$\left. + \frac{1}{2}\xi^2 \left[\frac{\partial^2 f(v,t)}{\partial v^2}\phi(v,\xi) + 2\frac{\partial f(v,t)}{\partial v}\frac{\partial \phi(v,\xi)}{\partial v} + f(v,t)\frac{\partial^2 \phi(v,\xi)}{dv^2} \right] \right\} d\xi$$

Hence

$$\tau\frac{\partial f(v,t)}{\partial t} = -\frac{\partial}{\partial v}\int \xi f(v,t)\,\phi(v,\xi)\,d\xi + \frac{1}{2}\frac{\partial^2}{\partial v^2}\int f(v,t)\,\xi^2\phi(v,\xi)\,d\xi \tag{12.77}$$

We define the moments of ξ by

$$\langle \xi(v)\rangle = \int_{-\infty}^{\infty} \xi\phi(v,\,\xi)\,d\xi = M_1\tau\ (\text{say})$$

$$\langle \xi^2(v)\rangle = \int_{-\infty}^{\infty} \xi^2\,\phi(v,\,\xi)\,d\xi = M_2\tau\ (\text{say}) \tag{12.78}$$

Substituting from Eq. (12.79) into Eq. (12.78), we get

$$\frac{\partial f}{\partial t} = -\frac{\partial}{\partial v}(M_1 f) + \frac{1}{2}\frac{\partial^2}{\partial v^2}(M_2 f) \tag{12.80}$$

This is a partial differential equation of the velocity distribution function involving the first and second moments of the velocity increments and is known as *Fokker-Planck equation* (one-dimensional). Similar equation is obtained for the distribution of any other variable subjected to random influences. The equation can easily be generalized to several generalized coordinates. The Fokker–Planck equation can be applied to a great variety of random processes, to calculate the probability density of a transition.

Particles starting with an arbitrary velocity distribution will eventually arrive at the equilibrium distribution. Approach to equilibrium is thus another aspect of the Fokker–Planck equation (see exercise 7).

NYQUIST FORMULA AND FLUCTUATION–DISSIPATION THEOREM

The Brownian movement of electrons in matter causes thermo-electrical fluctuations. In electronic devices, these fluctuations, i.e. irregular changes in currents and voltages create noise. The effect was first observed by Johnson in 1927 and is known as *Johnson noise* or *thermal noise*. The explanation was given by Nyquist and is known as *Nyquist formula* (or *theorem*).

Consider a Brownian particle under the influence of an external force F as well as the fluctuating force $\beta(t)$. The Langevin equation is

$$m\frac{d^2x}{dt^2} + \delta\frac{dx}{dt} = F + \beta(t) \tag{12.81}$$

The electrical analogy is a series LCR circuit, the charge Q plays the role of the coordinate x and $\varepsilon(t)$, (say) represents the fluctuating voltage. The corresponding Langevin equation is

$$L\ddot{Q}(t) + R\dot{Q}(t) = -\frac{Q(t)}{C} + \mathscr{E}(t) \tag{12.82}$$

Let us consider the simplest system described by the Langevin equation (inertia of the particle is neglected, i.e. $m\ddot{x}$ is neglected)

$$\delta \dot{x}(t) = \beta(t) \tag{12.83}$$

which corresponds to

$$RJ(t) = \mathscr{E}(t) \tag{12.84}$$

the motion of a weightless Brownian particle, not subjected to any external field, i.e. the current flowing through an ohmic resistance. We have from Wiener–Khinchin theorem

$$C(\tau) = \langle \beta(t)\, \beta(t+\tau) \rangle = \int_{-\infty}^{\infty} \langle G(\omega) \rangle \, e^{i\omega\tau} \, d\tau \tag{12.85}$$

where $\langle G(\omega) \rangle$ is the spectral density of the random force $\beta(t)$. From Eq. (12.83), we get

$$x(t) = \frac{1}{\delta} \int_0^t \beta(t)\, dt \tag{12.86}$$

using Einstein relation given by Eq. (12.56), which gives

$$\langle x^2 \rangle = \frac{1}{\delta^2} \int_0^t \int_0^t \langle \beta(\theta)\, \beta(\theta') \rangle \, d\theta \, d\theta' = 2Dt \tag{12.87}$$

Differentiation of Eq. (12.87) with respect to t gives

$$\frac{1}{\delta^2} \int_0^t [\langle \beta(t)\, \beta(\theta) \rangle + \langle \beta(\theta)\, \beta(t) \rangle] d\theta = 2D = \frac{2kT}{\delta} \tag{12.88}$$

We replace the variable θ by $\tau = \theta - t$ in the first term and by $\tau = t - \theta$ in the second term, to get

$$\int_{-t}^{t} C(\tau)\, d\tau = 2kT\delta \tag{12.89}$$

Since the microscopic equations are invariant under time reversal, the correlation function must be a symmetric function of time, i.e.

$$C(\tau) = C(-\tau) \tag{12.90}$$

Hence from Eq. (12.90), we can write

$$C(\tau) = 2kT\delta\, \delta(\tau) \tag{12.91}$$

where $\delta(\tau)$ is the Dirac-delta function. Expressing it in the form of Fourier integral

$$\delta(\tau) = \frac{1}{2\pi} \int_{-\infty}^{\infty} e^{i\omega\tau} \, d\tau \tag{12.92}$$

and substituting from Eqs. (12.91) and (12.92) into Eq. (12.85), we get

$$\langle G(\omega) \rangle = \frac{kT\delta}{\pi} \tag{12.93}$$

which is the Nyquist formula for a Brownian particle.

For a charge Q flowing through a resistor R, the corresponding spectral density $\langle |\mathscr{E}|^2 \rangle$ of the random electromotive force $\mathscr{E}(t)$, we have in place of Eq. (12.92):

$$\langle |\mathscr{E}|^2 \rangle = \frac{kTR}{\pi} \tag{12.94}$$

The Einstein formula takes the form

$$\langle Q^2(t) \rangle = \frac{2kT}{R}\, t \tag{12.95}$$

In technical books, often $\mathscr{E}_\omega^2$ is used in place of $|\mathscr{E}|^2$. $\mathscr{E}_\omega^2$ is the effective mean square of the electromotive force taken only for positive frequencies. This gives $\mathscr{E}_\omega^2 = 2|\mathscr{E}|^2$ and the Nyquist formula takes the form

$$\mathscr{E}_\omega^2\, d\omega = \frac{2kTR}{\pi}\, d\omega \tag{12.96}$$

Alternatively,

$$\mathscr{E}_\nu^2\, d\nu = 4kTR\, d\nu \tag{12.97}$$

which gives the mean square voltage across a resistor R for a frequency range $d\nu$.

The thermal fluctuations of the electromotive force in a resistor give rise to the *white noise*, i.e. an irregular mixture of harmonic vibrations of all possible frequencies and with equal average amplitudes.

The generalization of Nyquist theorem is the fluctuation-dissipation theorem, which is valid for both classical as well as quantal systems.

Generalizing to an arbitrary linear system, we have the generalized impedance

$$Z(\omega) = \beta_\omega / v_\omega; \quad R(\omega) = \text{Re}\, Z(\omega) \tag{12.98}$$

where β_ω and v_ω are the Fourier components of the generalized force and the generalized velocity. Obviously, $v_\omega = i\omega Q_\omega$. We then have

$$\frac{1}{2}\langle |\beta_\omega|^2 \rangle = \frac{2kT}{\pi} R(\omega), \quad \frac{1}{2}\langle |v_\omega|^2 \rangle = \frac{2kT}{\pi} \frac{R(\omega)}{|Z(\omega)|^2}, \quad \frac{1}{2}\langle |Q_\omega|^2 \rangle = \frac{2kT}{\pi\omega^2} \frac{R(\omega)}{|Z(\omega)|^2} \tag{12.99}$$

For quantal systems, the mean energy

$$E(\omega, t) = \frac{\hbar\omega}{2} + \frac{\hbar\omega}{e^{\hbar\omega/kT} - 1}$$

of a quantal oscillator should replace kT of the classical expression:

$$\frac{1}{2}\langle |\beta_\omega|^2 \rangle = \frac{2}{\pi} E(\omega, T)\, R(\omega); \quad \frac{1}{2}\langle |v_\omega|^2 \rangle = \frac{2}{\pi} E(\omega, T) \frac{R(\omega)}{|Z(\omega)|^2}$$

$$\frac{1}{2}\langle |Q_\omega|^2 \rangle = \frac{2}{\pi} E(\omega, T) \frac{R(\omega)}{\omega^2 |Z(\omega)|^2} \tag{12.100}$$

The first of these relations is known as the fluctuation dissipation theorem as it connects the spectral density of the equilibrium fluctuations $\langle|\beta_\omega|^2\rangle$ with the non-equilibrium dissipative properties of the system (the generalized coefficient of friction $R(\omega)$). The fluctuations affect the motion through the mechanism of energy dissipation.

IRREVERSIBLE TRANSPORT PHENOMENA AND ONSAGER PRINCIPLE

So far, we have discussed only the equilibrium states or some fluctuations about the equilibrium states. We now consider briefly the theory of irreversible transport phenomena. Ohm law of electrical conduction, Fourier law of thermal conduction, Fick law of particle current density, etc. are examples of the irreversible transport processes. The relationship among irreversible transport flows are taken to be linear. Onsager (1931) established the principle of symmetry of kinetic coefficients (transport coefficients), which is also known as *Onsager reciprocity relation.* According to this principle, the transport coefficients are symmetric tensors, i.e.

$$A_{ik} = A_{ki} \tag{12.101}$$

and if there is an external magnetic field or Coriolis forces, then

$$A_{ik}(\vec{B}, \vec{\Omega}) = A_{ki}(-\vec{B}, -\vec{\Omega}) \tag{12.102}$$

This principle explains a lot of experimentally observed facts. For example, it explains the Kelvin's deduction of relations between Seebeck, Peltier and Thomson effects.

The Onsager principle is obtained in three stages. We first calculate the correlations among the fluctuations. In the second stage, the symmetry of ensemble averages is obtained through the microscopic reversibility of time. In the third stage, we obtain the symmetry restriction (the Onsager principle) by the microscopic linearity of the laws of the decay of fluctuations.

In Chapter 5, we discussed the probability of deviation of a single thermodynamic quantity. We now turn to the probability of simultaneous deviation of several thermodynamic quantities represented by $x_1, x_2, \ldots, x_n$. The entropy $S(x_1, \ldots, x_n)$ is then defined as a function of these quantities and the probability has a relation similar to Eq. (5.1). The entropy is maximum at equilibrium and is less than the maximum value at any instant of time at which the system is away from equilibrium. Expanding S in powers of x_i, we have

$$S = S_{\max} + \sum_i \left.\frac{\partial S}{\partial x_i}\right|_0 x_i + \frac{1}{2}\sum_i \sum_k \left.\frac{\partial^2 S}{\partial x_i\, \partial x_k}\right|_0 x_i\, x_k + \cdots \tag{12.103}$$

Since S is a maximum at equilibrium, $\partial S/\partial x_i = 0$ and $\partial^2 S/\partial x_i \partial x_k$ is negative. Taking only up to second order terms, we get

$$\frac{S - S_{\max}}{k} = -\frac{1}{2}\sum_{i,k} \beta_{ik} x_i x_k \tag{12.104}$$

with $\beta_{ik} = \beta_{ki}$.

The probability distribution is given by

$$w = \text{constant } e^{-(1/2)\sum_{i,k} \beta_{ik} x_i x_k} \tag{12.105}$$

The constant is found by the normalization condition. Equation (12.105) is the Gaussian distribution for more than one variable.

We now define the quantities X_i, thermodynamically conjugate to the x_i by

$$X_i = -\frac{\partial S/k}{\partial x_i} = \sum_k \beta_{ik} x_k \tag{12.106}$$

and note that

$$-\frac{dS}{k} = X_i\, dx_i = x_k\, dX_k$$

and

$$x_i = \sum_k \beta^{ik} X_k \tag{12.107}$$

where β^{ik} is the reciprocal matrix to β_{ik}.

We now evaluate the ensemble average of these quantities

$$\begin{aligned} \langle x_i X_k \rangle &= \int_{-\infty}^{\infty} \cdots \int x_i X_k w \, dx_1 \ldots dx_n \\ &= -\int \cdots \int x_i \left(\frac{\partial w}{\partial x_k} \right) dx_1 \ldots dx_n \end{aligned}$$

Integrating by parts and recalling that w vanishes for $x_i = \pm\infty$, we get

$$\langle x_i X_k \rangle = \delta_{ik} \tag{12.108}$$

where δ_{ik} is Kronecker delta.

Further with the aid of Eq. (12.106), we get

$$\langle x_i x_k \rangle = \beta^{ik} \tag{12.109}$$

and

$$\langle X_i X_k \rangle = \beta_{ik} \tag{12.110}$$

In the second stage, we use the principle of invariance of microscopic equations of motion under time reversal. If $x_1, \ldots, x_n$ occur at the time t, then the ensemble average of x_k taken at a time $(t + \tau)$ is expected to be the same as the ensemble average of x_k taken at a time $(t - \tau)$, i.e.

$$\langle x_k(t+\tau) \rangle = \langle x_k(t-\tau) \rangle$$

Multiplying both sides by $x_i(t)$ and taking the average, we have

$$\langle x_i(t)\, x_k(t+\tau) \rangle = \langle x_i(t) x_k(t-\tau) \rangle$$

Taken over a long period of time, the time average is same as ensemble average; hence they do not depend upon the origin of time. We can replace t by $(t + \tau)$ on the right-hand side to get

$$\langle x_i(t)\, x_k(t+\tau) \rangle = \langle x_i(t+\tau) x_k(t) \rangle$$

Subtracting $x_i(t)x_k(t)$ from both side and dividing by τ, we get

$$\left\langle x_i(t)\frac{x_k(t+\tau)-x_k(t)}{\tau}\right\rangle = \left\langle \frac{x_i(t+\tau)-x_i(t)}{\tau}x_k(t)\right\rangle$$

or

$$\langle x_i(t)\dot{x}_k(t)\rangle = \langle \dot{x}_i(t)\, x_k(t)\rangle \tag{12.111}$$

If an external magnetic field $\vec{B}$ or angular velocity $\vec{\Omega}$ is present, they must be reversed on the right-hand side of the equalities.

In the third stage, we shall assume that the decay of fluctuations follows the phenomenological macroscopic laws with linear form of relations. This follows from the fact that the rate of decay of fluctuation at every instant is entirely defined by the value of the fluctuation at that instant of time. Thus, for small fluctuations, the rate of decay can be expanded in terms of fluctuations and retaining only the linear term, we can write

$$\dot{x}_i = -\sum_j \lambda_{ij}x_j \tag{12.112}$$

and the fluxes are

$$J_i = \dot{x}_i = -\sum_k A_{ik}X_k \tag{12.113}$$

Obviously,

$$A_{ik} = \sum_j \lambda_{ij}\beta^{jk}$$

Substitution of Eq. (12.113) in Eq. (12.111) gives

$$\left\langle x_i\sum_j A_{kj}X_j\right\rangle = \left\langle \sum_j A_{ij}X_j x_k\right\rangle$$

Use of Eq. (12.108) yields

$$\sum_j A_{kj}\delta_{ij} = \sum_j A_{ij}\delta_{jk}$$

or

$$A_{ki} = A_{ik} \tag{12.114}$$

The symmetry of transport coefficients in the presence of external magnetic field or angular velocity holds under time reversal only when their sign is simultaneously changed, i.e. if the transport coefficients depend on $\vec{B}$ or $\vec{\Omega}$, we have

$$A_{ik}(\vec{B}) = A_{ki}(-\vec{B}), \quad A_{ik}(\vec{\Omega}) = A_{ki}(-\Omega) \tag{12.115}$$

The above is the Onsager principle of symmetry. This principle has initiated considerable work in the thermodynamics of irreversible processes.

LINEAR RESPONSE THEORY

If a time-dependent perturbation is applied to a system, which drives it away from equilibrium or keeps it in a steady-state non-equilibrium condition, then the response of the system is linearly related to the perturbation and can be expressed in terms of correlation function. For this reason it is called *linear response theory*, developed by Knbo (1957).

Let a system be initially at equilibrium with Hamiltonian H_0. Let a time-dependent perturbation $H'(t) = -\alpha(p, q)F(t)$ be added to it. Let us consider the steady state after the transients have died down. The Hamiltonian of the system is

$$H = H_0 + H'(t) = H_0 - \alpha(p, q)\, F(t) \tag{12.116}$$

Since the system was initially at equilibrium, we have $F(-\infty) = 0$. The time development of the distribution function is described by Liouville equation

$$\frac{df}{dt} = \frac{\partial f}{\partial t} + \sum_k \left(\frac{\partial H}{\partial p_k}\frac{\partial f}{\partial q_k} - \frac{\partial H}{\partial q_k}\frac{\partial f}{\partial p_k} \right) = 0 \tag{12.117}$$

or in terms of Poisson bracket

$$\frac{df}{dt} = \frac{\partial f}{\partial t} + \{f, H\} = 0 \tag{12.118}$$

The initial state at equilibrium is described by the distribution function, assuming canonical ensemble by

$$f_0 = Ce^{-H_0/kT} \tag{12.119}$$

If we assume that the external field is small, we can write the distribution function as sum of the unperturbed part and the small perturbed part

$$f = f_0 + \Delta f \tag{12.120}$$

Substituting this in Eq. (12.118) and using

$$\frac{\partial f_0}{\partial t} = -\{f_0, H_0\} \tag{12.121}$$

we get, up to terms in first order of the perturbation

$$\frac{\partial \Delta f}{\partial t} = -\{\Delta f, H_0\} - \{\alpha, f_0\}\, F(t) \tag{12.122}$$

In terms of Liouville operator (which is Hermitian), Liouville equation is

$$i\frac{\partial f}{\partial t} = \hat{L}f \tag{12.123}$$

Evidently,

$$\hat{L} = -i\sum_k \left(\frac{\partial H}{\partial p_k}\frac{\partial}{\partial q_k} - \frac{\partial H}{\partial q_k}\frac{\partial}{\partial p_k} \right) \tag{12.124}$$

Solution of Eq. (12.123) is

$$f(t) = e^{-i\hat{L}t} f_0 \tag{12.125}$$

Let A be any dynamical variable depending on $p(t)$ and $q(t)$ only, we have from Eq. (12.118)

$$\frac{dA}{dt} = \{A, H\} = i\hat{L}A$$

which on integration, yields

$$A(t) = A(0)\, e^{i\hat{L}t} \tag{12.126}$$

The expectation value of α weighted according to ensemble density

$$\langle \alpha(t)\rangle = \langle \alpha(0) | f(t)\rangle = \langle \alpha(0) | e^{-i\hat{L}t} f_0 \rangle = \langle e^{i\hat{L}t} \alpha(0) | f_0 \rangle = \langle \alpha(t) | f_0 \rangle \tag{12.127}$$

Since $\hat{L}$ is Hermitian and hence $e^{\pm it\hat{L}}$ is unitary.

Equation (12.127) shows that the expectation value α at a time t can be calculated either by following the evolution of ensemble density or the evolution of the dynamical function.

For the equilibrium state, with the aid of Eq. (12.119) we can show that

$$\{f_0, A\} = \frac{f_0}{kT}\{A, H_0\} = \frac{1}{kT} f_0 \dot{A} \tag{12.128}$$

In terms of Liouville operator, Eq. (12.122) can be written as

$$\frac{\partial \Delta f}{\partial t} = -i\hat{L}_0 \Delta f - \{\alpha, f_0\}\, F(t) \tag{12.129}$$

This is a linear first-order differential equation, which can be solved by the use of an integrating factor. The solution is

$$\Delta f = -\int_0^{\infty} e^{i(t'-t)\hat{L}_0} \{\alpha(t'), f_0(t')\}\, F(t')\, dt' \tag{12.130}$$

If we are interested in the change of the average value of some dynamical variable β from its value in the equilibrium state, i.e. the change due to the application of the external force, we have

$$\begin{aligned}
\langle \beta(t)\rangle &= \langle \beta(0) | \Delta f(t)\rangle \\
&= -\int_0^{\infty} F(t')\, dt' \int \beta(0) \exp[i(t'-t)\hat{L}_0]\{\alpha(t'), f_0\}\, dp\, dq \\
&= -\int_0^{\infty} F(t')\, dt' \int \exp[-i(t'-t)\hat{L}_0]\, \beta(0)\{\alpha(t'), f_0\}\, dp\, dq \\
&= \frac{1}{kT}\int_0^{\infty} dt'\, F(t') \int \beta(t-t')\, f_0\, \dot{\alpha}(t')\, dp\, dq
\end{aligned}$$

i.e.

$$\langle \beta(t) \rangle = \frac{1}{kT} \int_0^\alpha dt' F(t')\, \phi(t - t') \tag{12.131}$$

where

$$\phi(t - t') = \frac{1}{kT} \langle \beta(t - t')\, \dot{\alpha}(t') \rangle \tag{12.132}$$

or by putting $t' = 0$, we get

$$\phi(t) = \frac{1}{kT} \langle \beta(t)\, \dot{\alpha}(o) \rangle \tag{12.133}$$

$\phi(t)$ is called the response function.

If the applied time-dependent force is

$$F(t) = Fe^{i\omega t} \tag{12.134}$$

we have for the response

$$\langle \beta(t) \rangle = \chi(\omega)\, Fe^{i\omega t} \tag{12.135}$$

where $\chi(\omega)$ is some generalized susceptibility given by

$$\chi(\omega) = \int_0^\infty \phi(t) e^{-i\omega t}\, dt \tag{12.136}$$

Further, since the angular brackets mean equilibrium ensemble average, we have the stationary condition

$$\langle \alpha(0)\, \beta(t) \rangle = \langle \alpha(s)\, \beta(t + s) \rangle$$

and hence

$$\frac{d}{ds} \langle \alpha(s)\, \beta(t + s) \rangle = \langle \dot{\alpha}(s)\, \beta(t + s) \rangle + \langle \alpha(s)\, \dot{\beta}(t + s) \rangle = 0$$

which gives

$$\langle \beta(t)\, \dot{\alpha}(0) \rangle = -\langle \dot{\beta}(t)\, \alpha(0) \rangle$$

and the response function has another form

$$\phi(t) = -\frac{1}{kT} \langle \dot{\beta}(t)\, \alpha(0) \rangle \tag{12.137}$$

The relations here (due to Kubo) have been derived for a classical ensemble but the relationships are preserved with minor modification for a quantal ensemble as well.

For quantal ensemble, the distribution function is replaced by density matrix (statistical matrix), the Poisson bracket by a commutator and the integration over phase space by the trace operation.

The quantum mechanical Liouville equation is

$$\frac{\partial \hat{w}}{\partial t} = \frac{1}{i\hbar} [\hat{H}, \hat{w}] = -i\hat{L}\hat{w} \tag{12.138}$$

where $\hat{w}$ is the statistical operator and $\hat{L}$ is Liouville operator.

The time evolution of the operator $\hat{\alpha}(p,q)$ (Heisenberg picture) is

$$\frac{d\hat{\alpha}}{dt} = -\frac{1}{i\hbar}[\hat{H}, \hat{\alpha}] = i\hat{L}\hat{\alpha} \tag{12.139}$$

The formal solutions are

$$\hat{w}(t) = e^{-it\hat{L}}\hat{w}(0)$$

and

$$\hat{\alpha}(t) = e^{it\hat{L}}\hat{\alpha}(0)$$

The expectation value of α is

$$\langle \hat{\alpha}\hat{w}(t)\rangle = Tr[\hat{\alpha}\hat{w}(t)] \tag{12.140}$$

The equation for $\Delta\hat{w}$ is

$$\frac{\partial \Delta\hat{w}}{\partial t} = -i\hat{L}_0\Delta\hat{w} - \frac{1}{i\hbar}[\hat{\alpha}, \hat{w}_0]F(t)$$

whose solution is

$$\Delta\hat{w} = -i\int_0^t e^{i(t'-t)\hat{L}_0}\frac{1}{i\hbar}[\hat{\alpha}, \hat{w}_0]F(t')\,dt'$$

The expectation value of β is

$$\begin{aligned}\langle\hat{\beta}(t)\rangle &= \langle\hat{\beta}(0)|\Delta\hat{w}\rangle = Tr[\hat{\beta}, \Delta\hat{w}(t)] \\ &= -\frac{1}{i\hbar}\int_0^t dt'\,F(t')\,Tr\left\{\hat{\beta}, e^{i(t'-t)\hat{L}_0}[\hat{\alpha}, \hat{w}_0]\right\} \\ &= -\frac{1}{i\hbar}\int_0^{\alpha} dt'\,F(t')Tr\{\hat{\beta}(t-t')[\hat{\alpha}(t'), \hat{w}_0(t')]\}\end{aligned} \tag{12.141}$$

The response function is

$$\phi(t) = \frac{1}{i\hbar}Tr\{[\hat{w}_0, \hat{\alpha}]\hat{\beta}(t)\} = \frac{1}{i\hbar}Tr\{\hat{w}_0[\hat{\alpha}, \hat{\beta}(t)]\} \tag{12.142}$$

since the trace of a product is invariant with respect to a cyclic permutation.

For dilute gases, the results are in agreement with those obtained from kinetic theory of gases but for condensed matter, the calculation of the response function is difficult. The linear response theory is useful in the calculation of transport parameters and in the examination of the general features of relaxation, resonance and non-equilibrium phenomena.

ELECTRICAL CONDUCTIVITY

The electrical conductivity is measured experimentally by applying an electric field across the system and measuring the electric current that flows in response to the applied field. Several set of measurements are made and the results are averaged. If the applied electric field is $\vec{E}$, the perturbation

$$H' = -\alpha F(t) = -\sum_k e_k \vec{r}_k \cdot \vec{E}(t) \tag{12.143}$$

where $\sum_k e_k \vec{r}_k$ is total dipole moment.

The resulting current

$$\langle \vec{J} \rangle = \left\langle \sum_k e_k \vec{v}_k \right\rangle \tag{12.144}$$

The variables of the particles are

$$\alpha = \sum_k e_k r_k$$

and

$$\beta = \sum_k e_k v_k \tag{12.145}$$

Evidently, we have $\dot{\alpha} = \beta$.

The conductivity is defined by the relation

$$\langle \vec{J}_\omega \rangle = \sigma(\omega) \vec{E}_\omega \tag{12.146}$$

where the subscript ω denotes their frequency dependent character.

We then have from Eq. (12.136)

$$\sigma(\omega) = \int_0^\infty e^{-i\omega t} \phi_{\beta\beta}(\tau) d\tau \tag{12.147}$$

where $\phi_{\beta\beta}(\tau)$ is given by Eq. (12.133) as

$$\phi_{\beta\beta}(\tau) = \frac{1}{kT} \langle J(0) J(\tau) \rangle \tag{12.148}$$

If the electric field applied is in a-direction, then

$$H' = -\sum_k e_k r_{ka} E_a \tag{12.149}$$

and the component of current in the b-direction

$$J_b = \sum_k e_k v_{kb} = \beta_b = \dot{\alpha}_b \tag{12.150}$$

This will give

$$\phi_{ba} = \frac{1}{kT} \langle J_a(0) J_b(\tau) \rangle \tag{12.151}$$

and the conductivity is given as a tensorial quantity

$$\sigma_{ba}(\omega) = \frac{1}{kT} \int_0^\infty e^{-i\omega\tau} \langle J_a(0) J_b(\tau) \rangle d\tau \tag{12.152}$$

a relation first derived by Kubo.

DIELECTRIC RELAXATION

Consider a non-conducting isotropic gas. Let $\vec{E}$ be the applied frequency dependent electric field and the induced dipole moment be $\sum_k \mu_k$. The resulting polarization $\vec{P}$ is related to the electric field applied by

$$\vec{P} = \chi(\omega)\vec{E} \tag{12.153}$$

We have

$$\vec{D} = \vec{E} + 4\pi\vec{P} \tag{12.154}$$

where the electric displacement $\vec{D} = \varepsilon\vec{E}$, ε being the dielectric constant, we have

$$\varepsilon(\omega) = 1 + 4\pi\chi(\omega) \tag{12.155}$$

Let the applied electric field be $E(t) = E_0 e^{i\omega t}$. Since the response function is real, $\chi(\omega)$ must be complex and hence the dielectric constant $\varepsilon(\omega)$ is also complex. Let us write

$$\varepsilon(\omega) = \varepsilon'(\omega) - i\varepsilon''(\omega) \tag{12.156}$$

where $\varepsilon'(\omega)$ is the real part and $\varepsilon''(\omega)$ is the imaginary part. Since the imaginary part represents the absorption by the system, it is often called the dielectric loss.

The perturbing term is

$$H' = -\sum_k \mu_k E(t) \tag{12.157}$$

The response function is given by

$$\phi(t) = \frac{1}{kT}\left\langle \sum_k \dot{\mu}_k(0)\, \mu_k(t) \right\rangle \tag{12.158}$$

We make a mathematical manipulation like this

$$\langle \dot{\mu}(0)\, \mu(t) \rangle = \langle \dot{\mu}(-t)\, \mu(0) \rangle = -\frac{d}{dt}\langle \mu(-t)\, \mu(0) \rangle = -\frac{d}{dt}\langle \mu(0)\, \mu(t) \rangle \tag{12.159}$$

where μ now stands for the total dipole moment. This gives

$$\phi(t) = -\frac{1}{kT}\frac{d}{dt}\langle \mu(0)\, \mu(t) \rangle \tag{12.160}$$

The Debye theory of dielectric behaviour is obtained by assuming an exponential decay of the kind.

$$\langle \mu(0)\, \mu(t) \rangle = \mu_0^2\, e^{-|t|/\tau} \tag{12.161}$$

where τ is called relaxation time.

Substituting from Eqs. (12.161) and (12.160) into Eq. (12.136) gives

$$\chi(\omega) = \frac{\mu_0^2}{kT\tau}\left[\int_0^\infty e^{-t/\tau}\cos\omega t\,dt - i\int_0^\infty e^{-t/\tau}\sin\omega t\,dt\right]$$

$$= \frac{\mu_0^2}{kT}\left[\frac{1}{1+\omega^2\tau^2} - \frac{i\omega\tau}{1+\omega^2\tau^2}\right] \tag{12.162}$$

This gives

$$\varepsilon'(\omega) = 1 + 4\pi\frac{\mu_0^2}{kT}\left(\frac{1}{1+\omega^2\tau^2}\right)$$

$$\varepsilon''(\omega) = 4\pi\frac{\mu_0^2}{kT}\left(\frac{\omega\tau}{1+\omega^2\tau^2}\right) \tag{12.163}$$

These are well-known Debye equations. $\varepsilon'(\omega)$ decreases from a fixed value $\varepsilon'(\omega)$ with increasing frequency, with most of the decreasing taking place around $\omega = 1/\tau$. $\varepsilon''(\omega)$ changes from a small value, shows a maximum at $\omega = 1/\tau$ and decreases to a small value again.

The plots of $\varepsilon'(\omega)$ and $\varepsilon''(\omega)$ against logarithmic wavelength are shown in Fig. 12.3 and Fig. 12.4.

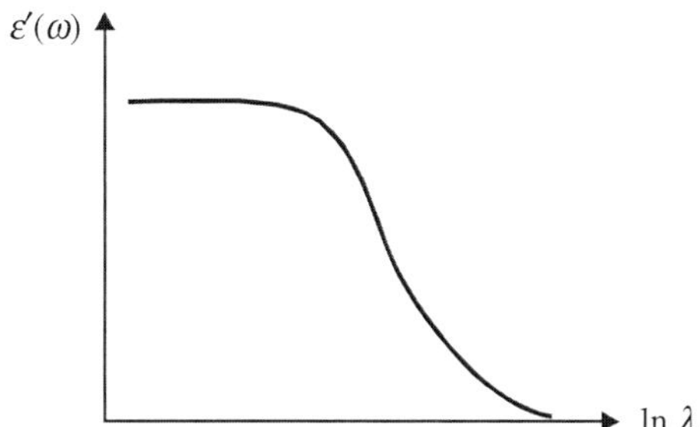

Figure 12.3 Variation of real part of dielectric constant with logarithmic wavelength.

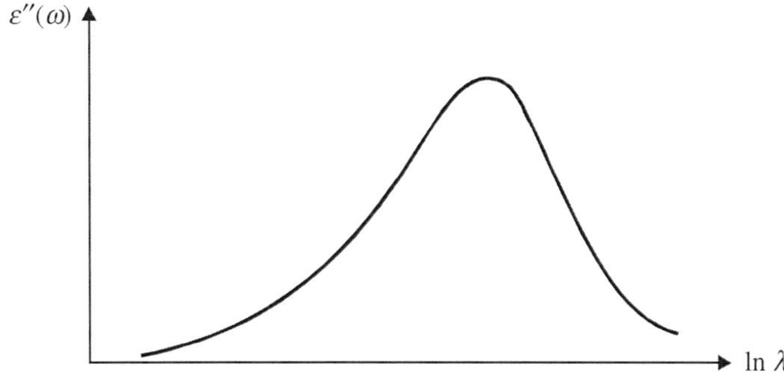

Figure 12.4 Variation of imaginary part of dielectric constant with logarithmic wavelength.

The agreement with experimental results in case of isobutyl bromide is very good.

A more general relation between $\chi'(\omega)$ and $\chi''(\omega)$, known as *Kramers–Kronig relations*, can be obtained as below.

We have

$$\chi(\omega) = \chi'(\omega) - i\chi''(\omega) = \int_0^\infty e^{i\omega t}\phi(t)\,dt \tag{12.164}$$

Hence

$$\chi'(\omega) = \int_0^\infty \phi(t)\cos\omega t\,dt, \quad \chi''(\omega) = \int_0^\infty \phi(t)\sin\omega t\,dt \tag{12.165}$$

Inversion gives

$$\phi(t) = \frac{2}{\pi}\int_0^\infty \chi'(\omega)\cos\omega t\,d\omega = \frac{2}{\pi}\int_0^\infty \chi''(\omega)\sin\omega t\,d\omega \tag{12.166}$$

Substitution for $\phi(t)$ from the second of Eq. (12.166) into the first of Eq. (12.165) yields:

$$\begin{aligned}
\chi'(\omega) &= \frac{2}{\pi}\int_0^\infty dt\cos\omega t\int_0^\infty \chi''(\omega')\sin\omega' t\,d\omega' \\
&= \frac{2}{\pi}\operatorname*{Lt}_{R\to\infty}\int_0^\infty d\omega'\,\chi''(\omega')\int_0^R \cos\omega t\sin\omega' t\,dt \\
&= \frac{2}{\pi}\operatorname*{Lt}_{R\to\infty}\int_0^\infty d\omega'\,\chi''(\omega')\frac{1}{2}\left[\frac{1-\cos(\omega'+\omega)R}{\omega'+\omega} + \frac{1+\cos(\omega'-\omega)R}{\omega'-\omega}\right]
\end{aligned}$$

The integrals containing the cosine terms will vanish since the integrand will oscillate infinitely rapidly as $R \to \infty$. This gives

$$\chi'(\omega) = \frac{2}{\pi}\int_0^\infty \chi''(\omega)\frac{\omega' d\omega'}{\omega'^2 - \omega^2} \tag{12.167}$$

A similar relation obtained by substituting for $\phi(t)$ from the first of Eq. (12.166) into the second of Eq. (12.165) gives

$$\chi''(\omega) = \frac{2}{\pi}\int_0^\infty \chi'(\omega)\frac{\omega' d\omega'}{\omega'^2 - \omega^2} \tag{12.168}$$

It can be easily seen that the Debye relations given by Eq. (12.163) satisfy the Kramers–Kronig relations given by Eqs. (12.167) and (12.168).

EXERCISES

1. Derive Boltzmann transport equation.
2. State and prove Boltzmann *H*-theorem. How is entropy related to *H*-function?
3. Show that steady-state solution of Boltzmann *H*-theorem corresponds to the Maxwellian distribution of velocities.

4. Explain the phenomenon of Brownian movement. Set up and solve Langevin equation to obtain the mean square displacement of the particles undergoing Brownian motion.
5. What is Markov process? Find the mean square displacement of a particle in the random process and use it to obtain the Einstein relation for mobility.
6. What is time-correlation function? How is it related to the power spectrum of the fluctuations?
7. Derive Fokker–Planck equation. Apply it to the phenomenon of Brownian motion and show that as time progresses, it reduces to the Maxwellian distribution of velocities.

   Hint: Langevin equation is

$$\frac{dv}{dt} = -\frac{\delta}{m} v + \beta(t)$$

For a small interval τ, we have

$$\xi = v(t+\tau) - v(t) = -\frac{\delta}{m} v\tau + \int_t^{t+\tau} \beta(t)\, d\tau = -\frac{\delta}{m} v\tau + G(\tau)$$

$\langle G(\tau)\rangle$ over several intervals of τ is zero, since $\beta(t)$ is a random variable.

$$\therefore \quad \langle \xi \rangle = -\frac{\delta}{m} v\tau$$

and

$$\langle \xi^2 \rangle = \frac{\delta^2}{m^2} v^2\tau^2 - 2\frac{\delta}{m} v\tau\langle G\rangle + \langle G^2\rangle = \frac{\delta^2}{m^2} v^2\tau^2 + \langle G^2\rangle = \langle G^2\rangle + O(\tau^2)$$

The general solution of the Langevin equation is

$$v(t) = v_0 e^{-(\delta/m)t} + \exp\left(-\frac{\delta}{m}t\right)\int_0^t \beta(t')\, e^{(\delta/m)t'}\, dt$$

Dividing the time interval $0 - t$ into n intervals of τ, i.e. $n\tau = t$, we have

$$v(t) - v_0 e^{-(\delta/m)t} = e^{-(\delta/m)t} \sum_{j=0}^{n-1} e^{(\delta_j/m)\tau} G_j(\tau)$$

$$\therefore \quad \langle (v(t) - v_0 e^{-(\delta/m)t})^2 \rangle = e^{-2(\delta/m)t} \langle G^2\rangle \sum_{j=0}^{n-1} e^{2(\delta/m)j\tau} = \langle G^2\rangle \frac{1 - e^{-2(\delta/m)t}}{e^{2(\delta/m)\tau} - 1}$$

Since $\tau \ll (\delta/m)^{-1}$ and $t \gg (\delta/m)^{-1}$, we have

$$\langle G^2\rangle = 2\frac{\delta}{m}\tau\langle v^2\rangle = \frac{2\delta kT\tau}{m^2}$$

and

$$\langle \xi^2\rangle = \langle G^2\rangle = \frac{2\delta kT\tau}{m^2}$$

The Fokker–Planck equation becomes

$$\frac{\partial f}{\partial t} = \frac{\delta}{m} f + \frac{\delta}{m} v \frac{\partial f}{\partial v} + kT \frac{\delta}{m^2} \frac{\partial^2 f}{\partial v^2}$$

Now introduce a new variable $\eta = ve^{(\delta/m)t}$, then

$$f(v, t) = f(\eta e^{-(\delta/m)t}, t) = \chi(\eta, t)\, e^{(\delta/m)t}$$

and

$$\frac{\partial f}{\partial v} = e^{(2\delta/m)t} \frac{\partial \chi}{\partial \eta}, \; \frac{\partial^2 f}{\partial v^2} = e^{(3\delta/m)t} \frac{\partial^2 \chi}{\partial \eta^2}$$

$$\frac{\partial f}{\partial t} = e^{(\delta/m)t} \frac{\partial \chi}{\partial t} + \frac{\delta}{m} v\, e^{(2\delta/m)t} \frac{\partial \chi}{\partial \eta} + \frac{\delta}{m} e^{(\delta/m)t} \chi$$

Substitution in Fokker–Planck equation of the above, gives

$$\frac{\partial \chi}{\partial t} = \frac{\delta kT}{m^2} e^{(3\delta/m)t} \frac{\partial^2 \chi}{\partial \eta^2}$$

Introducing now a variable $\theta = [(m/2\delta)(e^{(2\delta/m)t} - 1)]$ gives

$$\frac{\partial \chi}{\partial \theta} = D_0 \frac{\partial^2 \chi}{\partial \eta^2} \quad \text{where } D_0 = \frac{\delta kT}{m^2}$$

This has the same form as the diffusion equation.

The initial distribution is $f(v, 0) = \delta(v - v_0)$, hence $\chi(\eta, \theta = 0) = \delta(\eta - v_0)$. The solution of the diffusion-like equation under these conditions is

$$\chi(\eta, \theta) = \frac{1}{\sqrt{4\pi D_0 \theta}} \exp\,[-(\eta - v_0)^2 / 4D_0\theta]$$

Reverting back to the original variables, we have

$$f(v,t) = \chi(\eta, \theta)\, e^{(\delta/m)t} = \left[\frac{m}{2\pi kT\,(1 - e^{-2(\delta/m)t})}\right]^{1/2} \exp\left[\frac{-m\,(v - v_0 e^{-(\delta/m)t})^2}{2kT(1 - e^{-(2\delta/m)t})}\right]$$

As time progresses, for $t >> m/\delta$ it becomes

$$f(v, t) = \left(\frac{m}{2\pi kT}\right)^{1/2} e^{-mv^2/2kT}$$

which is the Maxwellian distribution.

8. Discuss the linear response theory and show that the response of a system in a steady-state non-equilibrium condition to a time-dependent perturbation can be expressed in terms of the time-correlation function.

9. Discuss Debye theory of dielectric relaxation.

10. Consider electron as a Brownian particle experiencing radiational friction, under the influence of thermal fluctuations in an electromagnetic field. Utilize the generalized Nyquist theorem to obtain Planck formula for the spectral energy density of the radiation.

Hint: Consider the electron of charge e and mass m moving in x-direction. The force of radiation reaction is $(2/3)\, e^2 \dddot{v}/c^3$. The Langevin equation (x component only) is

$$m\dot{v}_x - \frac{2e^2}{3c^3}\ddot{v}_x = e\mathsf{E}_x(t)$$

where E_x is the x component of the fluctuating electromagnetic field strength. For spectral components, we then have

$$\left(im\omega + \frac{2e^2}{3c^3}\omega^2 \right) v_\omega = e\mathsf{E}_{x\omega}$$

$$Z(\omega) = \frac{f_\omega}{v_\omega} = im\omega + \frac{2e^2}{3c^3}\omega^2 \text{ and } R(\omega) = \frac{2e^2}{3c^3}\omega^2$$

The spectral density of the energy of the electromagnetic field is

$$u(\omega) = \frac{V}{8\pi}\left(\frac{1}{2}\langle|\vec{\mathsf{E}}_\omega|^2\rangle\right) + \frac{1}{2}\langle|\vec{B}_\omega|^2\rangle = \frac{V}{8\pi}\langle|\vec{\mathsf{E}}_\omega|^2\rangle = \frac{3V}{8\pi}|\mathsf{E}_{x\omega}|^2$$

All the three directions being equivalent,

$$\langle|\vec{\mathsf{E}}_\omega|^2\rangle = 3\langle|\mathsf{E}_{x\omega}|^2\rangle$$

From Nyquist formula, we have

$$\frac{1}{2}\langle|\mathsf{E}_{x\omega}|^2\rangle = \frac{1}{2e^2}\langle|f_\omega|^2\rangle = \frac{2}{\pi e^2}E(\omega,T)\,R(\omega)$$

$E(\omega, T)$ is the mean energy of the quantal oscillator. This gives

$$u(\omega) = \frac{V\hbar\omega^3}{2\pi^2c^3} + \frac{V\hbar\omega^3}{\pi^2c^3}\frac{1}{e^{\hbar\omega/kT} - 1}$$

The first term on right-hand side corresponds to zero energy vibrations and the second term is the Planck formula.

Appendix

IMPORTANT PHYSICAL CONSTANTS

Quantity	Value	Uncertainty in the last two digits
Atomic mass unit (u)	$1.66053886 \times 10^{-27}$ kg	28
Avogadro constant	6.0221415×10^{23} mol^{-1}	10
Bohr magneton	$9.27400949 \times 10^{-26}$ J T^{-1} $5.788381804 \times 10^{-5}$ eV T^{-1}	80 39
Boltzmann constant	$1.3806505 \times 10^{-23}$ J K^{-1}	24
Electron rest mass	$9.1093826 \times 10^{-31}$ kg $5.4857990945 \times 10^{-4}$ u	16 24
Elementary charge	$1.60217653 \times 10^{-19}$ C	14
Gravitational constant	6.6742×10^{-11} N m^2 kg^{-2}	10
Molar gas constant	8.314472 J mol^{-1} K^{-1}	15
Molar volume of ideal gas (at 273.15 K and 101325 Pa)	22.413996×10^{-3} m^3 mol^{-1}	39
Neutron rest mass	$1.6749286 \times 10^{-27}$ kg 1.00866491560 u	10 55
Permeability of vacuum (μ_0) (magnetic constant)	$4\pi \times 10^{-7}$ $= 12.566370614... \times 10^{-7}$ N A^{-2}	Exact
Permittivity of vacuum (ε_0) (electric constant)	$1/\mu_0 c^2$ $= 8.854187817... \times 10^{-12}$ F m^{-1}	Exact
Planck constant (h)	$6.6260693 \times 10^{-34}$ J s	11
Planck constant over 2π ($\hbar$)	$1.05457168 \times 10^{-34}$ J s	18

Quantity	Value	Uncertainty in the last two digits
Proton rest mass	$1.67262171 \times 10^{-27}$ kg	29
	1.00727646688 u	13
Speed of light in vacuum	2.99792458×10^{8} m s^{-1}	Exact
Stefan–Boltzmann constant	5.670400×10^{-8} W m^{-2} K^{-4}	40
Wien displacement law constant	2.8977685×10^{-3} m K	51

IMPORTANT CONVERSION FACTORS

1 inch = 2.540 cm

1 metre (m) = 10^2 cm = 39.37 inch

1 angstrom (Å) = 10^{-8} cm = 10^{-10} m

1 kg = 10^3 g = 2.205 pound (lb)

1 atomic mass unit (amu) = 1 u = 1.6604×10^{-27} kg

1 gram per cubic centimetre = 1 g cm^{-3} = 10^3 kg m^{-3}

1 newton (N) = 10^5 dyne

1 joule (J) = 10^7 erg = 0.2389 calorie (cal)

1 kilo-calorie (kcal) = 4186 J

1 electron-volt (eV) = 1.6022×10^{-19} J

1 pascal (Pa) = 1 N m^{-2} = 10 dyne cm^{-2} = 0.01 millibar (mbar)

1 bar = 0.9869 atm = 10^5 Pa

1 atm = 101325 Pa = 760 mm Hg

1 tesla (T) = 10^4 gauss

1 weber (Wb) = 10^8 maxwell

References

Agarwal, B.K. and M. Eisner, Wiley Eastern, New Delhi, 1988 (Revised Edition, New Age International Publishers, New Delhi, 1998).

Andrews, F.C., *Equilibrium Statistical Mechanics*, John Wiley, New York, 1963.

Atkins, K.R., *Liquid Helium*, Cambridge University Press, Cambridge, 1939.

Chisholm, J.S.R. and A.H. de Borde, *An Introduction to Statistical Mechanics*, Pergamon, New York, 1958.

Feenberg, E., *Theory of Quantum Fluids*, Academic Press, New York, 1969.

Feynman, R.P., *Statistical Mechanics*, W.A. Benjamin, London, 1972.

Gibbs, J.W., *Elementary Principles in Statistical Mechanics*, Yale University Press, New Haven, 1902.

Gopal, E.S.R., *Statistical Mechanics and Properties of Matter*, Ellis Horwood, Chichester, Sussex, 1974.

Hill, T.L., *Statistical Mechanics*, McGraw-Hill, New York, 1956.

Huang, K., *Statistical Mechanics*, John Wiley, New York, 1963.

Khalatnikev, I.M., *An Introduction to the Theory of Superfluidity*, W.A. Benjamin, New York, 1965.

Kittel, C., *Elementary Statistical Physics*, John Wiley, New York, 1968.

Kubo, R., *Statistical Mechanics*, Interscience Publishers, New York, 1965.

Landau, L.D. and E.M. Lifshitz, *Statistical Physics*, Pergamon, London, 1958.

London, F., *Superfluids*, Vol. II, John Wiley, New York, 1954.

Ma, S.K., *Statistical Mechanics*, World Scientific, Philadelphia, 1985.

Mayer, J.E. and M.G. Mayer, *Statistical Mechanics*, John Wiley, New York, 1940.

McQuarrie, D.A., *Statistical Mechanics*, Harper & Row, New York, 1976.

Reif, F., *Fundamentals of Statistical and Thermal Physics*, McGraw-Hill, New York, 1965.

Sommerfeld, A., *Thermodynamics and Statistical Mechanics*, Academic Press, New York, 1956.

Ter Haar, D., *Elements of Statistical Mechanics*, Rinehart, New York, 1954.

Terletskii, Ya P., *Statistical Physics*, North Holland Publishing Company, Amsterdam, London, 1971.

Tolman, R.C., *The Principles of Statistical Mechanics*, Oxford University Press, Oxford, 1938.

Index